THE
DAY
THE SUN
ROSE TWICE

THE
DAY
THE SUN
ROSE TWICE

THE STORY OF THE TRINITY SITE
NUCLEAR EXPLOSION
JULY 16, 1945

Ferenc Morton Szasz

University of New Mexico Press
Albuquerque

Library of Congress Cataloging in Publication Data

Szasz, Ferenc Morton, 1940–
The day the sun rose twice.

Bibliography: p.
Includes index.
1. Atomic bomb—New Mexico—Los Alamos—History.
2. United States. Army. Corps of
Engineers. Manhattan Engineer District.
3. Los Alamos (N.M.)—Description.
I. Title.
II. Title: Trinity Site nuclear explosion, July 16, 1945.
QC773.A1S93   1984   623.4'5119   84-7587

Library of Congress Catalog Card Number 84-7587

ISBN-13: 978-0-8263-0768-2
ISBN-10: 0-8263-0768-X

21 20 19 18 17 16   10 11 12 13 14 15

Once again to the three generations:
Mary and Ferenc P. Szasz
Margaret
Eric, Chris, and Maria

Contents

Preface

Herman Melville once wrote, "To produce a mighty book, you must choose a mighty theme." The judgments on the book will be left to the reader, but I have no doubts about the power of the theme. This is the story of the world's first nuclear explosion. It occurred at 5:30 A.M. on July 16, 1945, about 120 miles from Albuquerque, New Mexico. Its legacy is much in evidence today.

As usual, I am indebted to a great number of people for their aid in the preparation of this study. To begin with, I would like to give special thanks to the archivists and librarians who so patiently replied to my endless stream of requests: Walter Bramlett, Tony Rivera, Bill Jack Rodgers, Elaine Ruhe, Dan Baca, and Alison Kerr of the Los Alamos National Laboratory, Los Alamos, New Mexico; Hedy Dunn of the Los Alamos Historical Society; Stanley Hordes and his staff at the New Mexico State Archives and Records Center, Santa Fe, New Mexico; Edward J. Reese of the Modern Military Headquarters Branch of the National Archives, Washington, D.C.; David Warringer and Richard V. Nutley of the Department of Energy, Coordination and Information Center, Las Vegas, Nevada; Mitch Tuchman and Bernard Galm of the UCLA Oral History Program and Anne Caiger of the UCLA University Library, University of California at Los Angeles; Elizabeth B. Mason of the Columbia Oral History Program, Columbia University, New York City; Frances M. Seeber of the Franklin Delano Roosevelt Library, Hyde Park, New York; Benedict K.

Zobrist of the Harry S Truman Library, Independence, Missouri; Richard L. Ray of the National Atomic Museum, Albuquerque; and Spencer Wilson, Professor of History at the New Mexico Institute of Technology, who also doubles as the head of the Socorro Historical Society. I would also like to offer a special word of appreciation to Dana Asbury of the University of New Mexico Press and to Dorothy Wonsmos, head of the Interlibrary Loan Division, Zimmerman Library, University of New Mexico.

Personal interviews provide the most rewarding aspect of research in modern American history. I would like to thank the following people for taking time to share their recollections with me: Herbert L. Anderson, Hans A. Bethe, Berlyn Brixner, Ted Brown, Norris Bradbury, Holm Bursum, Jr., Roy W. Carlson, Sam P. Davalos, Neil Dilley, Frank DiLuzio, Hymer L. Friedell, Robert W. Henderson, Louis Hempelmann, Jr., Jack M. Hubbard, McAllister Hull, Jr., Leo M. Jercinovic, Kermit H. Larson, Dorothy McKibbin, John Magee, George Marchi, Lilli Marjon, James F. Nolan, Frank Oppenheimer, Lois Page, Thomas Treat, Stanislaw M. Ulam, Marvin and Ruby Wilkening.

Mel Merritt, Al Solomon, and Charles Wood always had valuable advice on whom to interview next. Louis Hempelmann, Jack M. Hubbard, Kermit Larson, John Magee, Noel Pugach, Jack Reed, Richard Robbins, Jr., and Donald Skabelund did yeoman work in giving a critical reading to various sections of the manuscript.

Kenneth T. Bainbridge critiqued an entire draft and provided valuable personal insights. I would also like to thank Kermit Larson for lending me his copies of the UCLA Trinity surveys, Mrs. Stafford Warren for allowing me to read her husband's reminiscences, Kenneth T. Bainbridge for letting me read his Columbia oral history, and Barton C. Hacker for lending me the draft version of his "Elements of Controversy: A History of Radiation Safety in the Nuclear Weapons Testing Program." I would like to credit L. George Moses and George Lubick of Northern Arizona University, Flagstaff, and Alfred Castle and William Gibbs of the New Mexico Military Institute in Roswell for permitting me to test out some of my early ideas on their unsuspecting students. Marion Honhart and, especially, Penelope Katson should be lauded for their efforts in deciphering my handwriting, some of which, I discovered, I couldn't read myself.

A special mention of thanks should go to Jack M. Hubbard, now retired in northern California. Hubbard graciously allowed me full access to his diary and his film record of the Trinity shot, and patiently answered all my questions about meteorology.

I also owe a word of appreciation to Eric, Chris, and Maria. They listened to a steady barrage of Los Alamos stories at dinner with only an occasional murmur of complaint. Chris even donated part of her Christmas vacation to help compile the bibliography. Finally, my thanks again to Margi, for her encouragement, her companionship, her mastery of the historical craft, and her sound editorial judgment. Her constancy in this regard is, as Shakespeare phrased it, "a wondrous excellence."

THE
DAY
THE SUN
ROSE TWICE

Introduction

In December of 1945, the Indian pueblo of San Ildefonso, New Mexico prepared to celebrate its annual deer dance. That year they invited a square-dance club from nearby Los Alamos to join them, for the citizens of the famous "Atomic City" were soon preparing to leave the area. The Anglos from the Hill supplied several cases of Coca Cola while the Indians provided posole, green chile stew, and other native dishes. After the guests performed several square-dance routines, the Indians responded with one of their ancient dances. Finally, to the unusual accompaniment of accordion and Indian drums, the two groups danced together. In the midst of the festivities, the governor of the pueblo climbed on a bench and shouted above the din. "This is the atomic age," he cried. "This is the atomic age."[1]

It was, indeed, the atomic age, and it had its beginnings on July 16, 1945, about 180 miles south of San Ildefonso in an area of the central New Mexico desert the Spanish termed *Jornada del Muerto*. The scientists called it Trinity Site, and this is how it appears on most modern maps.

The name *Trinity* is not well known by the general public. It evokes no image of recognition as do, say, *Gettysburg, Pearl Harbor,* or *Harpers Ferry.* A book entitled *The Story of Trinity* might well be misfiled in the religion section of many bookstores.

Yet New Mexico's "Trinity" falls into a far different category. As the site for the world's first successful atomic detonation, it threw open doors that can never again be closed. What happened

3

at Trinity that Monday morning must go down as one of the most significant events in the last thousand years.

Surprisingly little has been written on Trinity. In 1946, William L. Laurence, distinguished science correspondent for the *New York Times*, wrote his book *Dawn Over Zero*. As Laurence was the only newsman to observe the blast, his version is the account of record. Nine years later, *Time* magazine reporter Lansing Lamont published his lively study *Day of Trinity*. Although the book is unfortunately marred by errors of fact, even J. Robert Oppenheimer, director of Los Alamos, admitted that Lamont captured the mood of the time with perfection.[2] In 1976, the official account by Harvard physicist Kenneth T. Bainbridge, who oversaw all the on-site operations, was declassified by the Los Alamos Scientific Laboratory. Bainbridge's encyclopedic pamphlet lists all the personnel and all the experiments involved in the test.[3]

But much remains to be told. By 1983, the federal government had declassified the majority of the relevant materials on the Manhattan Engineer District, or Manhattan Project (the code name for America's atomic bomb program). Numerous participants have also published their memoirs or diaries. With a perspective of almost forty years, it is time again to reexamine the story of the Trinity Site explosion.

The blast occurred in a stretch of the high New Mexico desert that was originally part of the Camino Real. This was the royal road north from Mexico City to the farthermost regions of Spanish settlement in upper New Mexico. By the early twentieth century, however, the area had been turned into ranch land, and thousands of sheep and cattle grazed the region as best they could. The state of New Mexico owned most of the land but leased it to the ranchers at a nominal cost.

With the outbreak of World War II, the federal government requested the area for use as a practice bombing range. Thousands of acres were leased to the government, with the assumption that they would be returned to the ranchers after the war was over. In 1944, a search team under Major General Leslie R. Groves, military head of the Manhattan Project, selected the region as the best site for the Trinity test. In about six months, the intense efforts of the Army Corps of Engineers, civilian contractors, and the Los Alamos scientists turned Trinity into a sprawling, open-air scientific laboratory.

The successful detonation on July 16, 1945, proved that the scientists' theories were accurate. It surpassed their wildest expectations. The effects of the blast were felt in three states, and nearby residents wondered aloud if the end of the world had arrived. The scientists were only slightly less awed by what they had created. It was an immensely pleased, yet somewhat uneasy, group of men who made their way back to Los Alamos in the early morning hours.

The success at Trinity had immediate international repercussions. General Groves flashed word to Washington, where it was relayed to President Harry Truman in Potsdam, Germany. With this information in hand, Truman set in motion the events that led to the bombing of Hiroshima and Nagasaki, the end of World War II, and (perhaps) the nuclear arms race with the Soviet Union.

The Trinity explosion left a local legacy as well. Within a few months, hundreds of cattle that had been foraging in the area of the blast developed unusual white blotches on their backs. Sensitive industrial X-ray film at the Eastman Kodak factory in Rochester, New York also developed strange spotting. Both were eventually traced to radioactive fallout from the Trinity explosion. In 1947, a special investigative team from UCLA arrived secretly for the first of several extensive studies of the effects of the radioactivity on the soil, plants, animal life, and people in the vicinity of the detonation.

From 1945 on, the National Park Service has periodically tried to make the Trinity area into a national monument, but their efforts have always been blocked. At first, the Atomic Energy Commission opposed the idea for reasons of health safety; later, the army opposed it for reasons of security. The land has not reverted to the area ranchers, for the old Alamogordo Bombing Range has now become the White Sands Missile Range. Today it forms one of the integral components of America's modern defense system. While the Park Service hopes to make Trinity Site into a national monument "eventually," no one cares to predict when that will be. The actual site of the detonation, lying in the northwest corner of White Sands Missile Range, is opened to the public only once a year, the first Saturday in October.

Everything connected with Trinity involved a "first," of course, but the scientists' success that morning proved especially significant in three areas: the Trinity test evoked the first serious analysis of whether or not an atomic explosion might somehow ignite the atmosphere and thus bring about the end of life on earth. It produced the world's first experience with airborne radioactive fallout and introduced the uncertainties that still exist in this area. Finally, the decisions Harry Truman made because of Trinity inaugurated a debate among scientists, historians, theologians, and now the general public, that shows little sign of disappearing. What follows, then, is the saga of Trinity Site. But first it is necessary to lay a little background.

CHAPTER ONE

The Origins
of Los Alamos

The late nineteenth–early twentieth century witnessed major conceptual revolutions in most areas of human knowledge. From the world of art came Impressionism, Cubism, Futurism, and Surrealism. These movements encouraged the collapse of the one-dimensional, "common sense" perspective that had been popular since the Renaissance. From the realm of music came new sounds of polytonality, atonality, and dissonance in general. Some composers even spoke of the "art of noises." Owing to the growing acceptance of germ theory and antisepsis, medicine moved from a folk art to a scientific discipline. From biology came the theory of genetics and a more sophisticated understanding of Charles Darwin's concept of evolution. Botany, geology, chemistry, history, economics—virtually all fields into which man had divided knowledge—underwent similar reorientations.[1]

Yet probably no discipline went through more of a conceptual revolution during those years than physics. Ever since the seventeenth century, the mechanical principles of Sir Isaac Newton's laws had dominated people's understanding of the physical universe. But in less than two generations, his laws lost their universality and their mechanical underpinnings as scientists began to piece together the picture of a relativistic and subatomic world. Gradually they created the now-familiar model, where each atom exists as a "miniature solar system" with negatively charged electrons rotating around a positively charged nucleus. Life in this subatomic world, moreover, seemed to follow its own laws. Experimental results were based less on actual observation than on "statistical probabilities."

Beginning with W. K. Roentgen's discovery of X-rays in 1895, major advances cascaded from the scientists' laboratories. In 1897, J. J. Thompson found the electron, and three years later Madame Eve Curie discovered radium. From 1902 on, Ernest Rutherford, "the father of the nuclear age," began a series of experiments that revealed what the atom was really like. Drawing upon experimental evidence, by 1913 Danish physicist Niels Bohr had imposed radical, inexplicable, constraints on Rutherford's theories. By the end of World War I, the Rutherford-Bohr model of the atom was generally accepted.

During the years 1900–1901, German physicist Max Planck arrived at his quantum theory of radiation. Beginning in 1905, Albert Einstein developed what is probably still the world's most famous equation: $E = mc^2$, where E is energy, m is mass, and c, the speed of light. During the 1920s, the Germans Wolfgang Pauli, Werner Heisenberg, Erwin Schrödinger, and Arnold Sommerfeld all made major contributions to the world of physics. Eugene Paul Wigner, who was a student in Germany during the 1920s, later recalled that "the whole of quantum physics was being created within my own eyesight."[2]

In 1930, Robert J. Van de Graaff built a generator to create a stream of subatomic particles, and the next year Ernest O. Lawrence and M. Stanley Livingston constructed the first cyclotron. But the most exciting discovery of all came in 1932. That year British physicist James Chadwick (later knighted) discovered the neutron. Because the neutron carried no electrical charge, it proved to be the ideal tool with which to bombard other elements. In 1934 in Paris, Jean Frederic Joliot-Curie and his wife, Irene, produced artificial radioactivity. In 1938, Italian physicist Enrico Fermi won the Nobel Prize for his experiments with slow neutron bombardment. It was the "age of personality" in physics, and it seemed as if the discoveries would never end.[3]

During those heady years, the small world of theoretical physics formed a genuine international community. The focus, however, lay in Europe. The great centers of research were Leyden, Cambridge, Copenhagen, Göttingen, Munich, Berlin, Rome, and (perhaps) California. Scientists crossed both national boundaries and oceans with ease. Nuclear physics in those early days, it was said, became simply "international gossip."[4]

But an equally important revolution was occurring in the realm of politics. The rise of Fascism in Hungary, Germany, and Italy soon began to affect the ivory-towered world of physics. The antisemitic laws of Hitler and Mussolini and the attacks on "Jewish physics" sent a stream of immigrants to Sweden, England, and the United States. Laura Fermi has estimated that between 22,000 and 25,000 of these people, in every field imaginable, ended up as permanent American residents.[5]

Perhaps one hundred of these émigrés were physicists. While academic jobs were not exactly plentiful in depression America, a surprising number were eventually hired by the American universities.[6] The most prominent of these immigrants were Albert Einstein, who was invited to join the Institute for Advanced Studies at Princeton, and Enrico Fermi, who left Italy after receiving the 1938 Nobel Prize to teach at Columbia. But other American schools also joined in. George Washington University hired Gregory Breit and Edward Teller; the University of Rochester, Eugene P. Wigner; Cornell, Hans Bethe; Johns Hopkins and, later, the University of Chicago, James Franck; Fordham, Victor Franz Hess; the Carnegie Institute of Technology, Otto Stern; and Stanford, Felix Bloch. The list could be extended.

Then, in December 1938, came the momentous discovery by German chemists Otto Hahn and Fritz Strassman. While bombarding uranium with neutrons, they discovered the unaccounted-for presence of barium in their experiment. (Barium is about half the atomic weight of uranium.) This puzzle was first unraveled by Lise Meitner, Hahn's long-term assistant, and her nephew, Otto Frisch, as a splitting of the uranium atom in two. Borrowing a term from biology, they called the process "fission." Meitner had been forced to flee to Sweden, and Frisch was visiting her in the small town of Kungälv near Göteborg when she disclosed her discovery. Frisch, who had been working in Niels Bohr's Copenhagen research laboratory, told his mentor, who, by chance, was preparing to leave for a physicists' conference at George Washington University, in Washington, D.C.[7]

On January 27, 1939, when Bohr announced this discovery to the conference, the physicists were astonished.[8] Several from the immediate area rushed to their laboratories, some still in full evening dress, to repeat the experiment. Within weeks, countless uranium atoms were fissioned by neutron bombardment. By the end of the year, over a hundred papers on the subject had broken into print in the various scholarly journals.[9]

Almost overnight, the enormous energy that had bound the atom together since the beginning of the universe had become available to man. The combined masses of the fission products totaled slightly less than the parent: the difference emerged as an enormous quantity of energy. The energy released per fission of a uranium nucleus approached 200 million electron volts. It was said that this fission of a single nucleus could make a grain of sand jump. Moreover, if, as expected, each nucleus released additional neutrons in the process of fission, then a self-sustaining reaction could occur. (The average emission of neutrons per fission turned out to be 2.5.) As soon as he heard about the Hahn-Meitner discovery, Enrico Fermi began speculating on the size of the crater that one kilogram of uranium would make if fissioned.[10] A pound of the fissionable isotope of uranium (U-235) was estimated to be the equivalent of 15,000 tons of TNT. In early 1939, the small community of physicists realized that a new, terrible weapon had become a real possibility.

Both the American-born scientists and the American government were slow to respond to this new situation. The early voices of warning about potential danger from the new weapon all came from the émigrés. This was largely due to the fact that in the United States the nation's universities seldom had close ties with the federal government. In general, the American government and the American universities moved in separate spheres that only occasionally overlapped.

Such was not the case in Europe. All of the great continental universities had strong connections with the state. Even the theological schools in Germany had governmental support— something unheard of in America. Thus, the émigré scientists knew well what use the emerging Fascist regimes might make of the new discoveries in the laboratories.

Theoretical physics suddenly began to have implications that extended far beyond the classroom. "It is still an unending source of surprise for me," recalled Polish mathematician Stanislaw M. Ulam, "to see how a few scribbles on a blackboard or on a sheet of paper could change the course of human affairs."[11] But such was the course of physics in the late 1930s. Up until 1938, theoretical physics ranked as one of the most esoteric of all disciplines. It was comparable, perhaps, to Medieval architecture or English Renaissance poetry. Overnight it became transformed into the most political of all fields.[12]

Thus it was that three Hungarian émigré scientists, Leo Szilard, Eugene P. Wigner, and Edward Teller, prevailed on fellow émigré Albert Einstein to write his now-famous letter to President Franklin Delano Roosevelt, warning him of the potential consequences of nuclear fission.[13] This letter, which Einstein publicly regretted after Hiroshima, was given to another émigré, presidential advisor Alexander Sachs, whom the scientists knew well.

Sachs had been friends with Roosevelt since the early 1930s, and while he had no official White House position, he was respected by the president. Sachs served as head of the Lehman Corporation and was well versed in economics, physics, music, and religion. On October 11, 1939, he gave Einstein's letter to Roosevelt. He feared, however, that the president would not comprehend the seriousness of it, so he asked for another visit to elaborate on its importance. He was told to return the next morning for breakfast.

Sachs later confessed that he spent a sleepless night in devising his plan of attack. When time for his breakfast appointment arrived, he told Roosevelt a story: Lord Acton, the noted English historian, was once asked if there were ever a time when England was saved, not through her own actions, but by the failure of an enemy to seize advantage of an opportunity. Acton asked for a day to think. When he responded, he said, yes, there was. Napoleon Bonaparte had tried to invade England but had failed because of the tricky tides and currents of the English Channel. Shortly afterwards, he was approached by the young American inventor Robert Fulton, who recommended the French build a fleet of steamships. Napoleon refused. Had he built them, Acton remarked, the whole of nineteenth-century history might have been different. After hearing the story, Roosevelt fell silent. Then he scribbled a note to the servant, who was clearing the dishes. The servant returned with a wrapped package, which turned out to be a magnum of Napoleon brandy. Sachs and Roosevelt clicked their glasses, and Roosevelt then told him that he would instruct his aides to follow through on this project.[14]

In spite of Roosevelt's promise to Sachs, however, relatively little was accomplished for almost two years. Roosevelt's initial "Advisory Committee on Uranium" was just that, and as late as April 1941, committee members expressed grave doubt that uranium research could contribute significantly to the war effort. It was not until October 11, 1941, that the president wrote to Winston Churchill to offer British nuclear physicists a plan whereby they could work in the United States.[15] The American military and political community had not yet learned to trust the émigré scientists. Their status as itinerant academics, many just in the process of attaining citizenship, provided few entrees to governmental circles, on any level. When Columbia University's Dean George H. Pegram arranged for Enrico Fermi to explain the possibilities of nuclear power to the navy, the officers were not impressed. "That wop talks crazy," one allegedly remarked.[16] The navy began a small program on nuclear energy, but it never assumed major proportions.

Consequently, the most important of the early responses to the splitting of the atom came not from the government but from the scientific community itself. Urged on by the farsighted, peripatetic, and irascible Leo Szilard (he actually took out a secret British patent on nuclear fission), the American physicists agreed among themselves to refrain from publishing any of their new discoveries. Their professional journals would accept the articles and write the necessary letters to deans and department chairmen. But they would print them only after the political climate had changed.

This decision to withhold publication turned out to have important, long-term consequences. As would be pointed out after the war, nuclear fission is a fact of the natural world. There is no "One Big Secret" that can be permanently hidden, for nature lies open to all trained investigators. But the nuclear world contains a great number of little, technical secrets that can be withheld successfully—at least temporarily. One of those was the finding of a proper "moderator" in which to encase the uranium and thus slow down the neutrons emitted in the fission process. It was crucial that this moderator absorb few neutrons itself. The two chief candidates were carbon and heavy water (a substance that contains oxygen and a rare form of hydrogen termed deuterium).

Working with Columbia physicist Herbert L. Anderson, Enrico Fermi had completed an article detailing how pure carbon could almost certainly be utilized as the perfect moderator. At Dean George Pegram's urging, he agreed not to publish it. Meanwhile a leading German scientist, Walter Bothe, conducted a fission experiment with an impure batch of carbon. The impurities absorbed many neutrons, and Bothe concluded (wrongly) that carbon would not suffice as a neutron moderator. Thus, the German scientists became dependent exclusively on heavy water. Heavy water always remained in short supply, however, and several British Commando raids on the major production plant in Rjukan, Norway made it even scarcer. This lack of a proper moderator proved to be a crucial factor in slowing down the German atomic program. Had Fermi published his article, and had the Germans started off down the right track, there is no telling what might have happened.[17]

After the Japanese attack on Pearl Harbor, however, America's atomic program began to expand more rapidly. There now existed the possiblity for direct military application of the new discoveries. Shortly afterwards, Roosevelt set up a National Defense Research Committee (NDRC), to mobilize science for war, and later, the Office of Scientific Research and Development (OSRD). The former was headed by physicist Vannevar Bush and the latter by Harvard President James B. Conant. The discovery of plutonium in 1940—an alternative route to a nuclear reaction—plus a series of optimistic reports from the British scientists, led by physicist Rudolph Peierls, strengthened the fledgling American efforts. By late 1941, the British were confident that a U-235 bomb could be developed before the hostilities were over.[18]

The most important step, however, occurred in June of 1942, when the nuclear program was turned over to the army. On September 17, 1942, Major General Leslie R. Groves, the man who had helped supervise the building of the Pentagon, was placed in charge of the Manhattan Engineer District. (The code name was chosen because the original headquarters were in New York.) Although Groves did not have the academic brilliance or technical training to match the people he had to oversee, he possessed an uncanny ability to learn from the discussions. Over

the years he proved his worth as an excellent, no-nonsense ad-
minstrator. His lively, albeit somewhat biased, account of the
project can be found in *Now It Can Be Told* (1963).[19] Although
his blunt personality and rotund shape made him a convenient
scapegoat for numerous personal gripes, it is doubtful that the
Manhattan Project could have succeeded as well as it did without
him. Brusque and abrasive though he was, Leslie Groves could
get things done.

Less than three months after Groves took over the project,
Enrico Fermi headed a team of scientists that inaugurated the
first controlled nuclear chain reaction. Borrowing a squash court
under the west stands of the unused University of Chicago foot-
ball stadium, Fermi and several burly students piled up 500 tons
of graphite amidst 50 tons of uranium and uranium oxide in a
matrix until the pile reached 48 layers. The graphite and uranium
were both machined into blocks about the size of a loaf of bread
for ease in handling, but it was still a messy, awkward operation
to arrange properly. It took a little over a month to build the
structure.[20]

The University of Chicago lies in the heart of the city, and
while there existed no danger of explosion, some scientists were
concerned about potential spread of radioactivity over the surrounding area. Consequently, extreme safety precautions were
taken. The pile contained special control rods of cadmium, a
high absorber of neutrons, and these could be dropped into place
at a moment's notice. Three scientists, dubbed the "suicide squad,"
stood at the top of the pile ready to dump buckets of a cadmium
solution on it, should there be any type of "runaway" reaction.
All went as scheduled, however, and at 2:30 on December 2,
1942, about forty people watched the world's first self-sustaining
nuclear reaction. When the pile had been shut down, Eugene P.
Wigner brought forth a bottle of imported Chianti, and all of the
spectators drank to the success of the experiment in paper cups.[21]
"The event was not spectacular," Fermi wrote in 1952, "no fuses
burned, no lights flashed. But to us it meant that release of
atomic energy on a large scale would be only a matter of time."[22]
The success at Chicago has been termed the halfway mark on
the road to the bomb.[23] For many of the scientists involved in
that momentous occasion, however, this first controlled release

of nuclear power formed *the* miracle of the Manhattan Project. The rest was simply "engineering."[24]

While the role of industrial engineering in the atomic program still remains to be written, it can hardly be dismissed so lightly. Theoretical speculations formed only one part of the nation's atomic effort. Actual industrial production of the fissionable materials proved to be quite another. Here the cooperation of the giants of American industry—Kellex, DuPont, Union Carbide, Chrysler, Eastman Kodak, Allis-Chalmers, General Electric, Westinghouse, Stone-Webster, and numerous others—proved crucial to its success. Under the combined efforts of the army, industry, and labor, gigantic facilities were established at Oak Ridge, Tennessee, and Hanford, Washington. Their tasks were to produce sufficient quantities of the fissionable U-235 and of plutonium for the bombs.[25] At the time the operations began, these materials existed only in microscopic quantities.

The technical problems involved were enormous. The fissionable isotope of uranium (U-235) is relatively rare, occurring in uranium at a ratio of 1 to 139. In other words, in every 140 pounds of uranium, one pound of U-235 exists, but it is scattered throughout so that initially the scientists found it virtually impossible to separate the two. In 1940, it was estimated that it would take twelve million years, at the then rate, to extract a pound of U-235.[26] The creation of plutonium by bombarding uranium 238 with neutrons proved only slightly less complex. That several pounds of each fissionable material were delivered in less than two years is a testimony to the genius of American technological and industrial efforts.

After the war, Groves praised the production efforts at Oak Ridge and Hanford in the highest terms. He stated that the hardest part of the project was producing a sufficient quantity of the fissionable material. Groves likened the task to a "manufacturer who tried to build an automobile full of watch machinery, with the precision that was required of watchmaking, and the knowledge that the failure of a single part would mean complete failure of the whole project."[27] The Los Alamos dimension of the work, he implied, was actually secondary.[28]

Groves, obviously, was exaggerating for effect. What went on at Los Alamos (or Site Y, as it was officially known) proved the cogwheel of the entire program. The secret atomic city in the mountains of northern New Mexico contained the *crème de la crème* of the Manhattan Project.

The Los Alamos site was created for two main reasons. First, the project needed a special weapons laboratory that would put the bomb together. Second, and probably more important, Groves found himself caught in his own massive security regulations. From the beginning, he had insisted that the people involved with the various aspects of the Manhattan Project know only enough to carry out their own jobs effectively. This "compartmentalization" of tasks lay at the heart of all Manhattan Project security. It proved so effective that no information ever reached German hands. It might have succeeded with the Soviet Union, too, had not a member of the British delegation to Los Alamos, Klaus Fuchs, passed along inside secrets to Soviet couriers.[29]

Essential as compartmentalization might have been for security purposes, it was a hindrance on the purely scientific level. The constant exchange of ideas and information became vital to the scientists as they encountered problem after problem, all of which were interconnected. Moreover, as the project grew, the logistics of getting the proper people to the proper places proved cumbersome. After the war, Leo Szilard complained bitterly that Groves's insistence on compartmentalization actually hindered the development of the bomb by eighteen months. Groves, however, always defended his position.[30]

In early 1942, Groves decided that the project needed to create a new, isolated site where the scientists could all come together and talk openly. So, in the summer of 1942, after a brief search, Groves, Major John Dudley, and the newly appointed head of this installation, California physicist J. Robert Oppenheimer, selected the region of Los Alamos, New Mexico. Oppenheimer had known and loved this area for years, for his family had had a ranch in the nearby Pecos Mountains. Here he was able to combine his two great loves—physics and New Mexico. Moving swiftly, the government took over the facilities of an exclusive boys' preparatory school and the lands of about twenty-six other area inhabitants. Much was already government owned and soon a total of about 9,000 acres was acquired for the war effort. Oppenheimer assumed that facilities would be needed to house perhaps thirty scientists and their families. The "realists" of the time argued that they would need room for at least 500. At the end of the war, close to 6,500 people were living on the Hill.

In early 1943, Mrs. Dorothy McKibbin, a widow in her mid-forties, was hired to run the front office at 109 East Palace Avenue in Santa Fe. Everything that went to the Hill passed through her doors. One of the unsung heroines of the Manhattan Project, Dorothy McKibbin directed lost scientists and nervous soldiers to the proper buses, while politely but firmly discouraging those who did not belong. When uninvited people arrived seeking work, she played dumb and sent them away. About twelve couples were married at her adobe home on the Old Santa Fe Trail, and her gracious charm smoothed many a ruffled feather. "It's been the most exciting job in the world," she confessed later.[31]

From 1943 to 1945, the tiny community of Los Alamos, New Mexico formed an unreal world, part mountain resort and part military base. Locally it was often termed "the Magic Mountain" or "Shangri-La." Those who came to the town after World War II frequently wished that they could have worked there earlier. For those who shared in the experience, remarked physicist I. I. Rabi, "it was their great moment."[32] It marked the people for life. Not surprisingly, a large number who lived and worked there during the war decided to return to the Santa Fe–Los Alamos region to retire.

The reasons for this atmosphere were many. To begin with, the physical environment could hardly be surpassed. At 7,400 feet, Los Alamos (the name means "the poplars") is surrounded by the towering peaks of the Sangre de Cristo mountain range, some reaching 13,000 feet. Abrupt mesas, the world's largest extinct volcano crater (the Valle Grande), numerous ancient Indian ruins, bustling modern Indian Pueblos, and tiny Spanish-American villages still speaking a seventeenth-century patois all lie within an hour's drive. The summers are cool and dry, and the winters offer both skiing and skating. Spring and fall provide ideal opportunities for hikes or horseback rides through the surrounding countryside. The area still is one of the most striking in the continental United States.[33]

More important than the scenery, however, was the deep sense of purpose that the group shared. The men and women at Los Alamos formed an international community that was engaged in a life-or-death struggle to beat the Germans to the secret of atomic power.[34] This goal gave the town its fierce intensity. In 1975, physicist Hans A. Bethe confessed that never, either before or after, had he worked as hard as he did during his years at Los Alamos. "It was one of the few times in my life," said another well-known physicist, "when I felt truly alive."[35]

Oppenheimer recruited many of the top personnel himself. His job was made easier by the fact that the scientists knew they would be applying their talents for the benefit of their country. They also knew that if they succeeded, they would become a part of history. The chief "rival" of the Los Alamos operation lay in the Radiation Laboratory of the Massachusetts Institute of Technology (MIT). But by the middle of 1943, all of the major types of radar had been fairly well developed, and this freed a number of MIT scientists to go to Los Alamos.[36] After some initial hesitation, recruitment snowballed, and by 1944 virtually every American physicist of importance was involved in the project. Some, however, were drawn into the work more by fate than by enthusiasm. "I worked on the bomb," one physicist confessed later, "because everybody I knew was doing it."[37]

Even those at Los Alamos who knew little about the activities behind the fenced-in "Tech Area," knew that they were producing something "that would help end the war." When the scientists returned from the Trinity explosion, a custodian rushed up to physicist Fred Reines, grabbed his hand, and said, "Well, we did it, didn't we." "Yes," said Reines, "we sure did."[38] This commitment pervaded all levels of society.

One must add that the men and women of Los Alamos were then young. The average age was around twenty-seven. Mathematician Stan M. Ulam worried because he was all of thirty-four. Physicist Kenneth T. Bainbridge was almost an elder statesman at forty. Many couples began their families while living on the mesa. Nearly one thousand babies were born in that small community from 1943 to 1949; those 208 born during the war had birth certificates listing the place as simply Box 1663, Sandoval County, Rural.

It is probably safe to say that never before in the history of the human race have so many brilliant minds been gathered together at one place. Visitors walking through the spacious Fuller Lodge at lunch might see four to five Nobel Prize winners dining at the same time. If they had been able to divine the future, they would have known that seven other men would also become Nobel laureates. A list of those who worked at Los Alamos from 1943 to 1945 reads like a page from *Who's Who* for the world of science.

J. Robert Oppenheimer. Although he seemed at first an unlikely choice for the top position—he had no prior administrative experience—few quarreled with the results. More than any other person, Oppenheimer shaped the atmosphere of the project. Unfortunately, his subsequent fall from grace has somewhat dampened the recognition previously given for his earlier administrative success. Yet Oppenheimer's magnetic personality and brilliant mind made Los Alamos hum. "A lesser man couldn't have done it," said Norris Bradbury, his successor as director. As historian Stephane Groueff has written: "Of all the atomic scientists, Oppenheimer became the most intriguing, the most adulated. His former students imitated him. Junior scientists admired him fervently, and young secretaries blushed in his presence."[39]

Niels Bohr and his son Aage. Bohr arrived at Los Alamos after a breathtaking escape from Nazi-occupied Denmark. While on the Hill, he contributed greatly to numerous theoretical discussions. A Nobel Prize winner, genially termed "the father of us all," Bohr conceived the "liquid drop" model of the nucleus. His son, Aage, who accompanied him, also later won the Nobel Prize. While at Los Alamos they were thinly disguised as "Nicholas and Jim Baker."[40]

Enrico Fermi. This Italian-born genius used the occasion of winning the 1938 Nobel Prize to flee Italy for the United States. Fermi was unique among the great physicists of the twentieth century in that he excelled in both experimental as well as theoretical physics. He may have been one of the last men who knew almost all of the physics of his day. In addition, Fermi's gentle personality and balanced outlook made him one of the few people of whom nobody had an ill word. As Sam K. Allison recalled after his death, "We may have seen his physical energy before, or his basic balance, simplicity and sincerity in life before, or even possibly his mental brilliance, but who in his lifetime has ever seen such qualities combined in one individual?"[41]

Victor F. Weisskopf.   While at Los Alamos, someone pinned a sign reading "The Los Alamos Oracle" over "Vikki's" door, in testimony to the numerous problems he helped solve. Later he went on to a distinguished teaching career at MIT and also served as director of the Geneva-based Centre European pour la Recherche Nucleaire (CERN).[42]

Emilio Segrè.   Another Italian immigrant, Segrè collaborated with Fermi in the discovery of slow neutrons and of technetium. Later he became Fermi's biographer. From 1943 to 1946, he served as group leader at Los Alamos, and in 1959 he shared a Nobel Prize for discovery of the antiproton.

Stanislaw M. Ulam.   A "pure" mathematician who joked about being loath to apply his ideas to the world of physics, Ulam made major contributions to numerous problems. After the war he worked closely with Edward Teller on the development of thermonuclear energy.[43] The Ulam-Teller theory is still one of the most closely guarded national secrets.

Edward Teller.   Teller was a Hungarian-born, controversial genius who will always be associated with the development of the H-bomb. Although he often insisted on having his own way at Los Alamos, Teller was also given some of the most complicated and important jobs to work on.[44]

I. I. Rabi.   Rabi spent most of his war years working at the Radar Laboratory at the Massachusetts Institute of Technology. But he often came to Los Alamos as a troubleshooter for Oppenheimer. He later also won the Nobel Prize.[45]

John von Neumann.   Oppenheimer once confessed that von Neumann was the brightest person he ever met, and there were few to dispute this claim. Inventor of the theory of games, Hungarian émigré von Neumann was also a leading figure in creating the first large-scale electronic computers. As one of his colleagues remarked, "If one listens to von Neumann, one understands how the human mind should work."[46]

Hans A. Bethe.   As head of the Theoretical Division at Los

Alamos, German émigré Bethe helped predict how the exploding ball of fire would behave. His strength lay in his ability to find simple and general ways to deal with complex problems. He taught all his career at Cornell and in 1967 was awarded the Nobel Prize for his theories on how the sun and stars shine. When interviewed by an Albuquerque reporter in 1981 about his prize, Bethe quipped "without my theories, the sun wouldn't shine."[47]

In his memoirs, Otto Frisch, a member of the British delegation to Los Alamos, noted that he always felt he could knock on any Los Alamos door and soon find himself in a stimulating conversation on poetry, science, art, or music.[48] Never had there been a community like it before; never would there be so again. It was America's Athenian world.

In addition to the already established scientists, from 1943 to 1945 Los Alamos also housed numerous young men then at the beginning of their scientific careers: physicist Marvin Wilkening, currently a dean at the New Mexico Institute of Technology in Socorro; radiologist Louis Hempelmann, now retired from the University of Rochester; physical chemist John Magee, now retired from Notre Dame, just to name a few. At the bottom of this rather illustrious roster was a carefully selected group of enlisted men, the Special Engineer Detachment (SEDs). Sent to Los Alamos because they had already had some scientific background—a few even had Ph.D.s—SEDs often made valuable contributions to solving the numerous problems at hand. A significant number of them also went on to scientific careers, such as McAllister Hull, Jr., now provost of the University of New Mexico, and Val L. Fitch, now professor of physics at Princeton. Our neighbors, remarked Jane Wilson, a teacher in the Los Alamos High School and wife of physicist Robert Wilson, included "the great, the near great, the never-to-be-great."[49]

While the Los Alamos scientists did not all fit the stereotype of the long-haired oddball, they did display their share of eccentricities. I. I. Rabi recalled Los Alamos as "a hotbed of prima donnas," and General Groves once claimed them "the largest collection of crackpots ever seen." Popular Chef George Marchi recalled that when he walked to the kitchen at 4:00 A.M. to bake the day's bread and rolls, he would frequently pass a scientist walking the other way, lost in thought over some theoretical problem.[50] Edward Teller seemed to enter and leave the Tech Area on his own schedule, much to the annoyance of others. He also relaxed by playing the piano at all hours of the night. Another scientist refused to pay the nominal fine for having left a top-secret report on his desk overnight. The report was totally wrong, he insisted, so he was doing a service by leaving it out where an enemy agent might pilfer it.[51]

Whenever they became ill, one group of scientists would go to the library, look up their symptoms, and then make their own diagnosis—much to the annoyance of the medical staff.[52] In 1944, Oppenheimer personally had to request several distinguished physicists to speak only English, not German or Italian, in public places, since they frequently lapsed into their native tongues on the streets of Santa Fe.[53] "The only thing typical about the men of science at Los Alamos," said Jane Wilson, "was that they were atypical. . . . Living among them would not have been such fun had many of them not been a little peculiar."[54]

Under normal circumstances, say a modern corporation or a university, such a battery of intellectual giants might soon be divided into rival factions. Indeed, after the war, the scientists' bitter personal quarrels became public property.[55] But it would be a mistake to read these later disagreements back into the 1943–45 period. Although there were inevitable clashes between strong personalities and some grumbling about the "cliques" that had formed by late 1944, major fallings-out were rare. A surprising number of reminiscences testify to this harmony. "But when one considers that we lived, day after day, year after year, closely packed together—aware of every detail of our neighbors' lives— even to what they were having for dinner every night," recalled Kathleen Mark, wife of Canadian mathematician J. Carson Mark, "one can't help but marvel that we enjoyed each other so much."[56]

Perhaps another reason why the community of Los Alamos could get along with each other so well was they had a common enemy: the military, with its innumerable rules and regulations. General Groves was frequently excoriated *in absentia,* and the Hill people were so rude and ungracious to Lieutenant Colonel Whitney Ashbridge that he suffered a heart attack and had to be transferred. Philosopher David Hawkins, brought in by Oppenheimer as a go-between for the scientists and the army field engineers, frequently found himself in the middle of the disputes. The tension was such that one official army report noted that "a considerable portion of the business was done at arm's length."[57]

The town nearest to Los Alamos was Santa Fe, thirty-five miles to the southeast. Naturally, the Santa Feans began to wonder about all the activity on the Hill. Numerous rumors began to circulate, and many of them were quite wide of the mark. Los Alamos was accused of being a poison gas factory, a plant for spaceships, a submarine base for the Russians, a whiskey mill, or a camp for dissident Republicans. One man heard of the many "doctors" there and concluded that it was a hospital for pregnant WACs. In late 1943, the talk became so absurd that Oppenheimer sent librarian Charlotte Serber and her physicist husband Robert to Santa Fe deliberately to spread the rumor that Los Alamos was really making an "electric rocket." That became the semiofficial response for many.[58]

When one considers how many people worked at Los Alamos itself, let alone the Manhattan Project in general, the army's success in concealing its purpose was phenomenal. It became, indeed, "the best kept secret of the war."[59] The list of people who were heavily involved, yet knew nothing, proved a lengthy one. Neither the R. E. McKee nor the M. M. Sundt companies, which built most of the town of Los Alamos, nor the Ted Brown Company, which built Trinity Site, knew what their projects were for. With few exceptions, most of the Los Alamos wives were uncertain what their husbands were doing in the Tech Area. (Groves had insisted that the wives not be told.) Los Alamos housewife Lilli Marjon recalled the shocked looks when a newly arrived wife asked innocently, "And what does your husband do?" She was not invited back.[60] Bernice Brode, Ruth Marshack, and Laura Fermi all confessed that they really understood the project only after the dropping of the bomb on Hiroshima.[61]

But security also worked the other way around. From 1941 to late 1944, American Intelligence was unable to gather a clear picture of the German atomic effort either. All that was known was that Germany had forbidden all exports of Czechoslovakian uranium and had enlisted about two hundred top physicists, chemists, and engineers for their project.[62] Rumors from newly arrived refugees rushed in to fill the gaps, and many of their predictions were decidedly gloomy. One émigré estimated that the Nazis would have the bomb by early 1944.[63] Since many of the Los Alamos scientists were Jewish, they had no illusions of what they could expect from a Nazi victory. It was from them that the others learned firsthand of the suffering of relatives and friends in the Old World.

Los Alamos thus found itself engaged in an all-out race with the Germans, and the stakes were high. In December 1941, right after Pearl Harbor, James B. Conant told his friend Harvey H. Bundy that the Germans could never win the war—unless they were ahead of America in the development of the atomic bomb.[64] Many of the scientists had taken their degrees from German universities, and even those who had not maintained respect for German science. "The pronouncement by Hitler that Nazi Germany would rule the world for the next thousand years strongly affected my resolution," noted Kenneth Bainbridge, "as it did so many others."[65] Vannevar Bush felt the same way. An ironic aspect of this can be seen in the incident where a young Los Alamos WAC was instructed to request over the newly installed public address system that "Werner Heisenberg" please report to the office. After two days, somebody took pity on her and explained that Nobel Prize winner Heisenberg was in Berlin heading the German atomic program.[66]

Yet, curiously, the military seemed to have little sense of the political motivation of the scientists. As far as they were concerned, their only task was to build an atomic bomb in the shortest possible time.[67]

The fate of the German atomic quest lay unknown until the Allied invasion of the continent on June 6, 1944. One component of the D-Day landing party was the secret "Alsos" (Greek for "Groves") mission, headed by Major Boris T. Pash. Dutch-born physicist Samuel Goudsmit accompanied Pash. This contingent had the specific task of discovering just how far the Germans had progressed in their atomic research. As the advancing troops overran university towns and captured Nazi scientists, Goudsmit combed through their documents and interviewed the prisoners—his "enemy colleagues," as he graciously called them. By December 1944, Goudsmit had given the War Department all his information. He concluded that the Germans did not have anything like an atomic bomb. Hitler, who never realized its significance, concentrated on the development of the V-1 and

V-2 rockets. Moreover, along the way the Germans had made a series of political and theoretical wrong turns in their search for atomic power. The heavy pounding of cities and industrial plants by Allied bombers also played a role in disrupting the German program. Werner Heisenberg argued later that the reluctance of the numerous pro-German but anti-Nazi physicists to pursue the project proved significant.[68] At any rate, in December 1944, the Germans were found to be at least two years behind the Americans.[69] The German atomic bomb program was a failure.

Ironically, however, when this information arrived at Los Alamos, it had almost no effect on the project. The surrender on May 8, 1945, had even less. In fact, Oppenheimer recalled that "I don't think there was any time when we worked harder at the speed-up than in the period after the German surrender and the actual combat use of the bomb."[70] By early 1945, the scientists were totally involved with the plans for a test of the almost-completed atomic device somewhere in the United States.

There were two separate types of atomic weapons in preparation. The first utlized the rare isotope U-235. It was originally nicknamed "Thin Man" (after Roosevelt), but later renamed "Little Boy" (for nobody) when technical changes shortened the proposed gun barrel. The second utilized plutonium and was nicknamed "Fat Man" (after Churchill) from the beginning.[71] The scientists were much more confident about the former than the latter. In the uranium bomb, one subcritical mass of U-235 is fired at another subcritical mass. When the two join, they become supercritical and ignite in a nuclear reaction. The scientists were so confident about their theories on the uranium bomb that they planned no field test for it. Indeed, this was the first bomb dropped on Hiroshima. By July of 1944, however, the scientists had also concluded that this relatively simple "gun assembly" method would not work for the plutonium bomb. The two subcritical pieces of plutonium could not be brought together fast enough to prevent a premature detonation.

Led by physicist Seth Neddermeyer and mathematician John von Neumann, the solution to the dilemma emerged as the theory of "implosion." Here a subcritical sphere of plutonium was surrounded by charges of heavy explosives, all carefully shaped as "lenses." When these were detonated, they focused the blast wave so as to compress the plutonium instantly into a super-

critical mass.[72] This was a much more complicated procedure,
however, and many people expressed considerable doubt that it
would work as planned.

From May to November 1944, there was a fierce debate at Los
Alamos as to whether the scientists should field test the plutonium bomb before actually dropping it. Harvard explosives
expert George B. Kistiakowsky and Oppenheimer both argued
for such a test, but initially Groves was opposed. He was afraid
that if the test failed, the precious plutonium would be scattered
all across the countryside by the detonation of the shaped explosive lenses. He lived in constant fear of facing congressional
investigating committees if the Manhattan Project did not suc-
ceed in time to help end the war.

Eventually Groves was persuaded. The plutonium production
at Hanford was increasing at such a rate that a field test would
cause little delay in time. Besides, he was told, if the untested
plutonium bomb were dropped and did not perform as expected,
the enemy would find themselves owners of a "gift" atomic
weapon. Finally, as the ultimate compromise, a gigantic, 214-
ton, cylinder-shaped tank (called "Jumbo") would be created to
house the bomb. If a nuclear explosion occurred, Jumbo would
be vaporized. If only the conventional explosives detonated, the
vessel would be strong enough to contain the plutonium for yet
another try.[73] With this as the agreement, the search for a test
site began in earnest.

CHAPTER TWO

The Construction and Naming of Trinity

Finding a suitable spot to test "the gadget," as the bomb came to be known, started in May of 1944. When the first search team of J. R. Oppenheimer, Kenneth T. Bainbridge, and two army majors became bogged down in the late New Mexico spring snow, it added a rather romantic aura to the whole process. Although numerous people went on the various site surveys, the brunt of the search was conducted by Major W. A. (Lex) Stevens, who was to be responsible for the construction of the camp, and physicist Kenneth T. Bainbridge. Bainbridge was to be in charge of all on-site operations, and from the spring of 1944 until the fall of 1945 he was totally occupied with site-related issues. The search teams pored over numerous maps and visited several possible locations. They also borrowed a seven-passenger C-45 and scouted much of the western United States from the air.

The requirements for the test site were strict. The location had to be relatively flat, both to minimize the effects of the terrain on the blast and to maximize the opportunities for the numerous optical experiments. The weather in the region had to be basically good. The site had to be isolated from any centers of population, yet close enough to Los Alamos to provide ease of movement for both men and material. (This became more and more crucial as the time for the test approached.) The potential ease of transportation for Jumbo also played a more important role than anyone was willing to admit. Finally, Harold

Ickes, the secretary of the interior, had decreed that no Indians should be displaced because of it.[1]

At one time, eight different site locations were under consideration: the Tularosa, New Mexico basin; the Jornada del Muerto region of central New Mexico; a desert training area near Rice, California; an island off the coast of southern California; the desert region south of Grants, New Mexico, near the malpais; the area southwest of Cuba, New Mexico and north of Thoreau; the barrier sand reef off the coast of south Texas; and the San Luis Valley region of southern Colorado.

The choice soon narrowed down to three: the malpais region near Grants, New Mexico; the desert training area north of Rice, California; and the Jornada. The first two, however, presented problems. The malpais lay close to Los Alamos, but the region was difficult to cross. Engineers feared they would have to blast out large sections of the ancient lava beds to provide enough roads. They were also concerned about potential difficulties in hauling Jumbo over the lava. The California desert area could accommodate Jumbo without difficulty, but it lay too far from Los Alamos for easy travel. Moreover, it was soon discovered that General George Patton had been using it as a training ground for his Africa Corps troops. Groves once termed Patton "the most disagreeable man I ever met" and absolutely refused even to talk to him about the area.[2] This left the Jornada. In August of 1944, the Governing Board at Los Alamos finally recommended this site, and Groves gave his approval. It was made official September 7, 1944, in the Colorado Springs office of Uzal G. Ent, commanding general of the Second Air Force (Army Air Force).[3]

The Jornada del Muerto is a ninety-mile stretch of high desert lying between present-day Socorro and El Paso. Ranging between 4,000 and 5,000 feet, it is bounded on the east by the San Andres Mountains and on the west by the San Mateos. Winter weather is cool and mild, but July and August temperatures reach well over one hundred degrees. The sandy soil supports only yucca and scattered vegetation.

The history of the Jornada is long and complex. An unnamed captain from Francisco Vasquez de Coronado's expedition probably explored part of it in 1541, but no European crossed it entirely until Don Juan de Oñate arrived in 1598. After that it became an integral part of the Camino Real, the main highway from Mexico City, through El Paso, to the region of the north-ernmost Spanish settlements. Several camping spots used by Spanish caravans may still be seen from the air.[4]

The early Spanish expeditions dreaded this stretch of the long journey north for two reasons. First, the land lay flat and open, and this invited hit-and-run attacks from the more mobile Apache Indians. Even more important was the lack of water. It is here that the Rio Grande bends sharply west and for about ninety miles cuts through steep canyons and abrupt gorges. From what is now Hatch, New Mexico, northward, the Spanish caravans could no longer follow the Rio Grande. Although the road across the Jornada proved about fifty miles shorter, it also meant three to four days without water.[5]

All of the early travelers crossed the region with trepidation. After a night without water, Oñate's May 1598 expedition spot-ted a dog with muddy feet, and followed it to a temporary water hole. Don Diego de Vargas survived his August 1692 journey through the Jornada only because heavy afternoon thundershow-ers allowed his men to replenish their water supply.[6] Santa Fe Trail pioneer Josiah Gregg described the area as entirely destitute of water and crossed it only during the late afternoon and night. In 1852, a U.S. Army surgeon also commented on its stiff, un-pleasant winds.[7]

The name *Jornada del Muerto* is translated, variously, as "Route of the Dead Man," or "Journey of Death." The fact that the Span-ish named the little settlement at the northern end *Socorro* (suc-cor or sustenance) showed how relieved they were to get through it.

By the early 1940s, however, the Jornada had been tamed. Sparsely settled ranches dotted the region, running their sheep and cattle over the barren landscape. At the eastern edge, the Sierra Oscuro peaks rose to elevations of over 8,000 feet. Their piñon and juniper provided shade for the grazing areas and relief from the summer heat.

The state of New Mexico owned most of the Jornada and leased a good portion of it to ranchers for their cattle and sheep. In 1942, immediately after Pearl Harbor, the federal government requested several hundred square miles to be used for test bomb-ing, and this became the Alamogordo Bombing Range. The New Mexico commissioner of public lands exchanged the land, on an acre-for-acre basis, for some federal land on the eastern side of the state.[8] Two years later, Secretary of War Henry L. Stimson wrote New Mexico's Governor John L. Dempsey that the War Department needed additional acreage. As the state's educational institutions depended heavily upon land revenues for their operation, Dempsey was reluctant to transfer any more land but eventually did so. Where the federal government planned to make permanent installations, it purchased the land outright. Where it planned only temporary structures, or where the land would not be permanently altered—as, for example, in a practice bombing range—the lands were simply leased. The presumption was that the leased lands would eventually revert to the ranchers after the war.[9]

One reason that General Groves approved the Jornada for the test was that most of the land was already in the hands of the federal government under the leasing arrangement. But when the aftermath of the Trinity test and the subsequent establishment of the White Sands Missile Range made it impossible to return the land to the people as promised, it inaugurated a controversy that has still not been settled. About ninety-seven families were involved, but the most vocal have been the several members of the McDonald clan.

The first McDonald entered the Jornada in the late nineteenth century and homesteaded there. His several sons continued the tradition until they were asked to move because of the war. The original army lease arrangements contained no provisions for purchase. Since the McDonalds' land was leased, not purchased, the government simply renewed the leases year after year, regardless of the protests of the original owners, who always maintained that they were promised they could return. "We asked 'em, and they said we were supposed to get it back," David McDonald sourly remarked in 1982, as he recalled his long quarrel with Uncle Sam. "I call it a land steal."[10]

For over forty years, David McDonald has fought the American government for the return of his ranch and some additional mining claims in the Oscuro Mountains. In 1977, he sent the army an eviction notice, declaring them trespassers. Then, on the morning of October 13, 1982, eighty-one-year-old David and his niece, Mary, "occupied" their old ranch, posting signs that read: "Deeded land—No trespassing" and "Road Closed to U.S. Army."

While army security watched nervously, negotiations began. Finally, after Senator Harrison Schmidt and Congressman Joe Skeen (both R-NM) flew into the range, the McDonalds left peaceably. The issue involved money, and the government promised to examine the books to see whether the McDonalds had been fairly paid for their property. There is no chance, however, that David McDonald will ever get his ranch back.[11]

The choice of the specific spot within the Jornada eventually fell to Kenneth T. Bainbridge. Much of the land was to be "borrowed" from the Alamogordo Air Base, and its commander, Colonel Roscoe Wriston, did not relish relinquishing an eighteen-by-twenty-four-square-mile section of his command. Reluctantly he consented to releasing a block in the northwest corner of the bombing range. Because of the prevailing wind patterns for the area, Bainbridge initially had hoped for the southwest corner of the range, but eventually he agreed to the former. While it was known that the Oscuro peaks had great local influence on the weather pattern, detailed maps, geological data, and weather information for the region were virtually nonexistent.[12]

On Sunday, September 17, 1944, Bainbridge and a few others visited the probable site of the test. As they approached the area at 5:30 P.M., seven B-17s flew over the well-marked portable targets (labelled B-4) and dropped their practice bombs for about fourteen minutes. "If we had arrived ten minutes later," Bainbridge wryly noted in his official report, "we would have been at B-4 and it might have been more exciting."[13] As the area lay barren and open, it did not take Bainbridge's group long to stake out the approximate location for Ground Zero and the three observation sites, North 10,000, West 10,000, and South 10,000, named for their distance in yards (5.68 miles) from the point of detonation.[14]

By November of 1944, the Trinity construction operation was in full swing. Captain Samuel P. Davalos and about one hundred men from the Army Corps of Engineers began the building, aided initially by the J. D. Leftwich Construction Company of Lubbock, Texas. "You just did the job required," Davalos recalled years later. "You didn't go ask questions or worry about money."[15] The Corps of Engineers borrowed tents and ten-foot CCC wood sections from Albuquerque for the first buildings, and later renovated the McDonald Ranch house, adding work benches, room dividers, and map areas. They also repaired and weather-stripped the ramshackle doors and windows in the never-ending battle to keep out the dust.[16] Although the corps tarpapered the walls and painted the roofs with aluminum paint, this did little to alleviate the over one-hundred-degree temperatures of June and July. Imported generators provided electricity.

Davalos visited the site about twice a week to supervise construction, and much of his early concern revolved around the water problem. Two old windmills provided about a gallon an hour, when the wind blew, but the water contained numerous chemicals and was very hard. There was so much gypsum in it the ranchers termed it "gyp water" and often used it as a purgative. Eventually, the corps ordered specially built softeners, which alleviated the problem somewhat, but the army trucks still hauled in Rio Grande water on a steady basis.

Drawing on their experience at Los Alamos, the Corps of Engineers set up several portable outbuildings and laid numerous concrete platforms. By January 1945, a permanent crew of twelve MPs occupied the site.[17] Four barracks, a mess hall and kitchen, a commissary, a vehicle repair station, a complete storehouse (termed *Fubar* for "fouled up beyond all recognition") and a latrine followed in succession. The initial plans of 1944 anticipated housing and feeding facilities for approximately 160 men.[18] By July of 1945, however, the number approached 300. On the weekend of the test there were 425 people present at Trinity.[19]

Construction demands were so extensive that it was not long before the Corps of Engineers needed civilian help. In December 1944, Colonel Ruben Cole, director of the corps for a five-state region, called in Ted Brown, a well-known Albuquerque contractor who headed one of the state's largest construction firms. He told Brown that the army had been impressed with his company's past performance record. (They had built several landing strips and had brought every job in on time.) Now, the corps said, it had an even more important job for him. They asked that Brown sign a blank contract, the figures to be filled in later. This unusual provision reflected not so much secrecy as the fact that the scientists had not fully decided on everything to be constructed. The job expanded as the months went on, and Brown's company did a good deal of incidental work as needed.

At first Brown hesitated. He was concerned that he would be unable to procure the necessary building materials. The army, however, told him not to worry. This job had top priority, and they promised to get him everything he needed. The next day the officers took Brown in a dune buggy to the site and gave him detailed instructions. He was to begin by building about forty miles of road, plus several heavy beam and earthen observation bunkers.

Starting in March of 1945, Brown and his crew of 200 New Mexicans worked dawn to dusk, seven days a week, for thirty straight days. After everyone had gone home for a brief rest, the pace began again. The company worked three such phases. Although the construction itself was not unusual, the pressure on them was. The project was "hotter than anything we had ever gotten hold of," Brown later recalled.[20] Brown's crews set up hundreds of T poles, about five and a half feet high, running from Ground Zero to the various shelter locations. When they returned after one rest period, they discovered that the army had strung them with low-slung wire, which could be reached easily from the ground. Two large herds of antelope roamed the region, and several people worried that they might run through and snap the drooping wires. But no instrument line ever had to be replaced because of the antelope. Although the herds raced across the region at full speed, they always ducked under the wires at just the right moment.[21]

In addition to the line poles, the Brown people also built three large earth and concrete bunkers and two towers. The bunkers utilized heavy timbers, twelve inches by twelve inches or sixteen by sixteen, and each had a concrete slab for a roof. Everything was then covered by several tons of dirt all around. The bunkers were not built for aesthetics, Brown recalled later; they were built for strength. The builders realized that such structures obviously had something to do with an explosive force.

Brown's company also helped erect two large towers. The first, which reached twenty-five feet high, was built from the same heavy wooden timbers as the bunkers. When they were asked to scrape out a 150-yard fire break around the tower, they recognized that this, too, was designated for heavy explosives. After they returned from one break, the men discovered that the tower had completely disappeared. It had been incinerated in the May

7, 1945, 100-ton (TNT) test shot. The other tower was steel. Fabricated in the east, they helped erect it by laying four large concrete pyramids, buried twenty to twenty-five feet into the earth, for the base. The steel tower proved easier to handle than the wooden tower. When it was finished, it looked like an "overgrown windmill."[22]

The most difficult task for Brown's crew, however, was to construct and maintain the roads. The original plan was simply to use dirt roads, but the dust clouds from them rose half a mile in the air, and the scientists detoured across the land whenever possible. "The soil wasn't very good," Brown recalled in a 1980 interview. "It was mostly gypsum sand. You can pack it down, let three trucks go over it and the fourth truck turns the road into three inches of flour. It didn't water very well. There was no local source for water so we trucked water in from the Rio Grande. We had twenty trucks coming in twenty-four hours a day."[23] As time for the test approached, the condition of the roads assumed high priority, for the scientists worried over a possible emergency evacuation from the test area.[24]

Fortunately, someone found a mica-gravel deposit in the vicinity of the steel tower, which they dug and spread about six inches deep on the roads. This they watered thoroughly and then primed with a coat of asphalt. Groves eventually ordered twenty-five miles of road treated in such fashion, at a cost of about $5,000 a mile.[25] The area immediately under the metal tower was treated in a similar manner. While this was not exactly "paving," it served its purpose for the duration.

Perhaps the most mysterious of Brown's construction projects involved the scraping of a twenty-five-mile haul road from Pope's Siding, a little-used siding on the Santa Fe Railroad, to Ground Zero. This road was built especially for Jumbo.

The story of Jumbo proved a troubled one from the start. To begin with, the design demands for the container were enormous. If the 5,300 pounds of high explosives (HE) surrounding the plutonium core ignited the atomic material, the vessel would disappear instantly. If it did not, the container had to be strong enough to contain the blast so that a special recovery team could enter and retrieve the costly plutonium. Recovery would have been an enormous task, of course, but the scientists felt it within the realm of possibility.

Los Alamos teams began by creating small-scale models of the container and exploding scaled-down HE charges within them. None of the initial designs withstood the detonations, however, and various pieces of them could soon be found on nearby Los Alamos mesas. In late 1944, post engineer Robert W. Henderson lured his Berkeley engineering professor, Roy W. Carlson, to Los Alamos and gave him the assignment.

Carlson changed the shape of the vessel from a sphere to a cylinder and suggested the idea of surrounding it with concrete. The peak pressure from the blast achieved such intensity that an enormous amount of mass was needed to contain it—but only momentarily. After the concrete had been pulverized into dust, the strength of the container's banded steel walls would take over. The walls would balloon just enough to hold the charge and would then return to near normal. Prototype scale models of the new design bore out Carlson's theories. When General Groves was informed of Carlson's success, he grumbled, "I thought that elephant was dead." Thus Jumbo got its name.[26]

While the scale models proved successful, constructing a full-sized Jumbo was quite another matter. Vannevar Bush had to use his personal connections with the officials at Babcock and Wilcox Steel Corporation of Barberton, Ohio before they agreed to attempt it. The size of the container taxed the company's heat-treating facilities to their utmost. When they finished, however, they had made the world's largest pressure vessel. Shaped like a gigantic thermos, it weighed 214 tons and had fifteen-inch walls of banded steel.

Hauling Jumbo from Ohio to Trinity Site proved equally taxing. The Eichleay Corporation of Pittsburgh, specialists in heavy moving, modified a 200-ton railway car—a flatcar with a recessed center. Then, wrapped in canvas and under armed guard, the heaviest single object ever moved by the railroads began its journey to New Mexico. The train took a circuitous route: to New Orleans, across the Mississippi through Texas, and through Clovis to Belen, New Mexico. From there the Santa Fe Railroad hauled it to Pope's Siding, south of Socorro. It rested there for several days on a special spur, built just for that purpose. Veteran railway engineer John M. Brown later remarked that Jumbo was the "dangest thing" that the Santa Fe Railroad had ever hauled.[27] Many an eyebrow was raised as it lumbered through Socorro, especially when the wind whipped up the canvas to reveal its size.

After several days at Pope's Siding, the Eichleay team transferred Jumbo onto a specially built sixty-four-wheel trailer. Then nine tractors hauled it slowly across the road to Ground Zero. It arrived in early April. By then, however, the scientists had grown much more confident about the expected results of the test, and there existed considerable doubt that Jumbo would be needed at all.

The new confidence rested on two factors: an improved set of high explosives and the development of the "implosion" lens system. Before World War II, the science of explosives remained virtually a black art. One could not accurately estimate the yield of a blast until after it had been ignited. Scientists at Los Alamos first had to design a variety of heavy explosives that would ignite simultaneously over the entire surface and also deliver the same amount of blast each time. Eventually, the family of explosives named "Composition B" began to show all the desired characteristics.

From his work with conventional explosives during the first part of the war, chemist George Kistiakowsky had concluded that detonation was a process which was "completely described by rigorous laws of thermo-dynamics and hydro-dynamics." Consequently, it could be totally controlled.[28] The scientists then began to shape the explosive charges to create a set of "lenses" to "focus" the blast wave. The resulting blast would then collapse the two plutonium hemispheres to form a supercritical mass. This replaced the earlier designs, which had involved attempts to collapse a thin plutonium shell. The new model proved much more reliable.

Moreover, several scientists argued that exploding the gadget inside Jumbo would distort all their complicated instrumentation—the *raison d'être* for the test. John Magee later confessed that he had long felt the whole idea preposterous. So, with the new confidence added to the old doubts, Oppenheimer and Bainbridge finally decided to bypass the gigantic vessel. "Jumbo," Bainbridge complained later, "was a very weighty albatross around our necks."[29]

Something had to be done with the container, however, so the scientists hastily erected a seventy-foot steel tower about half a mile from Ground Zero and placed Jumbo on a concrete base within it. No scientific measurements were planned for the arrangement, however, and the experiment was, at best, half-hearted. The ensuing blast collapsed the tower, but Jumbo survived undamaged.

Actually, the decision to bypass Jumbo proved wise. Had the container been utilized, it would have been instantly vaporized under the force of the nuclear explosion. Not only would it have thrown off all the test measurements, it would also have added 214 tons of radioactive steel particles to the mushroom cloud. These would have eventually returned to New Mexico as fallout.

In the meantime, security around Trinity remained tight. Lieutenant Howard C. Bush and his MP detachment maintained a permanent station at the site from January of 1945. From there, they patrolled the region on horseback and by jeep. Soon they were joined by Counter-intelligence officer John Harold Anderson, and he became the chief security liaison for Trinity. At first, officials hoped that no one would associate Trinity with Los Alamos. Several of the itinerant construction workers of the area, however, immediately recognized Bush and some of the other men as being from the Hill. At times, various people took trips to nearby El Paso and Ciudad Juárez, Mexico, bringing back Mexican jewelry and artifacts for their wives in Los Alamos.[30] An alert observer might have noticed this.

Security remained foremost in the minds of the men involved. Whenever they spoke to one another over the radio, they used only each other's initials. Whenever meteorologist Jack Hubbard confronted a mechanical breakdown at Trinity, his first thought was usually sabotage.[31] While the official correspondence occasionally fretted about possible security leaks, there were none of importance.[32] Moreover, anyone with a minimum of intelligence could have established that there was a connection between Los Alamos and the new construction site at Trinity.

Almost as important as security, however, was keeping a high morale among the men stationed there. The region was bleak at best, and in June and July it approached the impossible. The gypsum soil tended to cake on sweaty backs and frequently caused skin irritations, broken only by dips in the cattle-watering reservoirs. All food at the mess hall had to be trucked in, and the two cooks were decidedly mediocre. Only an occasional fresh antelope steak livened up the standard army diet.[33]

A major "gripe session" was held, and problems of food, transportation, guard schedules, and so on received a thorough airing. In March, the test director recommended that better recreation facilities be established for the MPs and engineers stationed there. They especially urged that more 16 mm movies be brought in.[34] Photographer Berlyn Brixner remembered outdoor movies every night of the week. Lieutenant Bush preferred adventure films, and so the men viewed such classics as *The Prisoner of Zenda* and *Beau Geste* several times. Yet, all things considered, Bush did an excellent job of keeping spirits high. Bainbridge always gave him the highest marks in this regard.[35]

As soon as the decision was made to test the bomb at Trinity, the scientists began to lay plans for their numerous experiments. The test data, it was argued, could provide valuable information for future weapons improvement, especially since it would be impossible to gather similar data from any military explosion over an enemy land. The effects of air blast, ground shock, neutron and gamma ray flux, the rise and spread of hot gasses, and the distribution of radioactive residue were the main items to be measured. Since the Trinity bomb would be exploded near the ground, this would also establish whether other such bombs should be detonated at or near the earth—to rely primarily on the force of earthquake and ground shock—or several thousand feet in the air.[36] The issue of what experiments should be performed—those essential versus those merely desirable—always proved a thorny one. The various last-minute suggestions for additional experiments bothered Bainbridge considerably. In later articles he recommended that no new tests be added after four weeks before any target date.[37]

By February 1945, General Groves had ordered an end to further scientific exploration at Los Alamos. From then on, all efforts were channeled toward the July target day.[38] A special Cowpuncher Committee was formed to "ride herd" on the test. Last-minute technical problems at the Los Alamos end regarding the delivery of the lens molds and various metallurgical difficulties proved especially annoying. On June 9, Bainbridge noted, "J. R. Oppenheimer has requested that Trinity *must not delay* the test of the gadget and we must schedule operations based on *the earliest possible date.*"[39] The earliest possible date had initially been July 4, but now was moved to July 13, at 4:00 A.M., weather permitting.

By early July, then, construction at Trinity had reached completion. A base camp housing 300 men, 500 miles of wires, several miles of roads, 25 of them black-topped, heavy earth and concrete bunkers, numerous smaller photographic bunkers and other non-manned stations, hundreds of gauges and instruments, a tower for Jumbo, elaborate radar and searchlight operations—all were ready to go.[40] Over fifty cameras, some mounted on converted machine gun turrets, were set up to record the behavior of the expected mushroom cloud and its eventual dispersion.[41] Los Alamos photographic specialists Julian Mack and Berlyn Brixner had devised special shielding devices and viewfinders for their cameras and placed them all through the area. The film footage, some sections only recently released to the public, proved to be startling.

Many of the scientific experiments were extremely delicate. All the devices near Zero had to relay their information the instant before they were incinerated. Because of these demands, the caliber of instrumentation was the most sophisticated the world had ever seen. It took years before such standards of instrumentation were reached again. The men involved worked without a sense of hours or of limits. The magnitude of the obstacles overcome, distances especially, was overwhelming. "It was a will to do," one observer recalled, "that has long since disappeared."[42]

The on-site test director, Kenneth Bainbridge, and his chief assistant, John Williams, maintained a firm rein on all this activity. Numerous reminiscences praise these two in glowing terms. Bainbridge, especially, was seen as a soft-spoken, quiet physicist who listened to the men and then got the best out of them, largely by force of personality.[43] He did "a most marvelous job of organizing the whole thing," recalled I. I. Rabi.[44]

This was not always easy, however, for it is clear from Bainbridge's accounts that the period immediately preceding the test was filled with intense excitement. The loudspeaker system connecting the three shelters remained in constant use, and for two weeks prior to July 16, the switchboard was manned on a twenty-four-hour basis. Snafus appeared with maddening regularity. Sergeant A. H. Jopp, a power company lineman in civilian life, and the only man who knew where the 500 miles of wire started and stopped, was side-swiped by a civilian-driven Mack truck. Although shaken up, he fortunately was not seriously hurt.[45]

Oppenheimer always remained concerned over the outdoor nature of the experiments. The scientists placed their instruments at great distances from Zero, and many wires lay in the open in the desert. Everything was made as fail-safe as possible, but no one was certain if a snapped cable or broken wire might not inaugurate something unexpected.[46]

Everything pointed to the test day in July, "the earliest possible moment." On-site rehearsals were made over and over again so that nothing was left to chance. The entire desert area had been turned into an outdoor scientific laboratory. Although no exact figures are available, Sam Davalos estimated that the test preparations cost approximately five million dollars.[47] At the center of this activity, on a 100-foot steel tower, rested "the gadget." It always remained under armed guard.

Why all this activity should have been labeled *Trinity* is a bit of a puzzle. Indeed, *Trinity* still seems a strange, almost blasphemous, term for such a spot. There is, moreover, no agreement on how the site received its name.

In 1982, Robert W. Henderson maintained that Colonel Lex Stevens named it. According to Henderson, he and Stevens were at the site surveying the best way to haul Jumbo from the railway siding to Ground Zero. A devout Roman Catholic, Stevens observed that the siding was "Pope's Siding." He remarked that the Pope had a special access to the Trinity, and that the scientists would need all the help they could get to move Jumbo to its proper spot. From this, the name *Trinity* gradually became common parlance.[48]

A more common version suggests that J. Robert Oppenheimer selected the name, but there is no agreement as to what he had in mind when he chose it. He, himself, never said. Physicist Robert Jungk thought that *Trinity* was taken from an old abandoned turquoise mine located in the region. Others have suggested that it was connected with the fact that three bombs—

the "unholy trinity"—were under construction at the same time. Reporter Lansing Lamont has argued that just before Oppenheimer was asked for a name, he had been relaxing by reading a John Donne poem: "Batter my heart, three person'd God; for you / As yet but knock, breathe, shine, and seek to mend . . ." Lamont claimed that Oppenheimer chose the name from this.[49]

Assuming that Oppenheimer did choose the name, these theories are still unlikely. In 1974, historian Marjorie Bell Chambers offered the most probable explanation. She suggested that Oppenheimer's choice of *Trinity* referred not to the Christian understanding of the term but to the Hindu connotation of it. Oppenheimer's religious background, of course, was Jewish, and his early schooling came from Felix Adler's Ethical Culture School in New York City. Since he had taught himself Sanskrit for pleasure, Oppenheimer was well conversant with Hindu culture. The Hindu concept of Trinity consists of Brahma, the Creator; Vishnu, the Preserver; and Shiva, the Destroyer. For Hindus, whatever exists in the universe is never destroyed. It is simply transformed. The cycle of life is such that if one part dies, another is created from it. It was in this sense, Chambers noted, that Oppenheimer chose the word *Trinity*.[50]

The name seemed a bit awkward, even from the first. During the early stages of planning, Bainbridge complained to Oppenheimer that the new *T* designation was becoming confused with the various other *T*'s that already existed. Because of this, some people were abbreviating the site as *A* and others as *R*. Oppenheimer standardized the designation as *TR*. This became official.[51]

The awkwardness has persisted. Although Trinity Site appears on most New Mexico maps, it is usually spoken of as "Trinity, near Alamogordo." Alamogordo, New Mexico, however, is over sixty miles away. The closest town of any size is Socorro, about thirty miles to the northwest of the site, but Socorro is seldom associated by the popular mind with the events there. Perhaps Trinity is best defined geographically: latitude 33° 28'–33° 50', longitude 106° 22'–106° 41'. Perhaps the ambiguity of the name is reflected in the ambiguity of the situation itself.

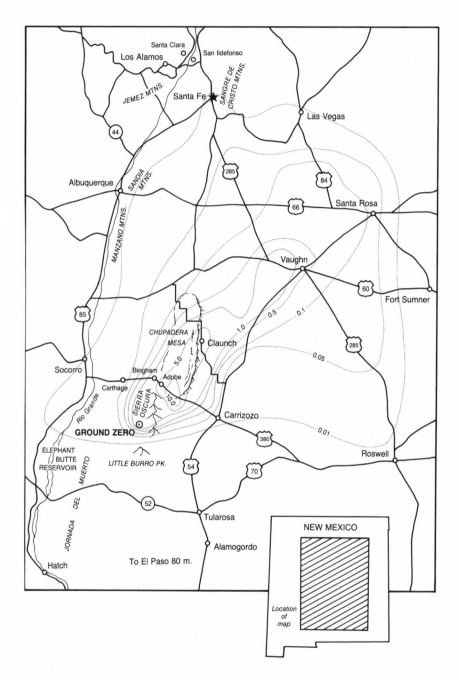

Central New Mexico. Gray lines show early measurements of fallout
pattern (in Roentgens). Data from LA 1027-DEL.

0.053 SEC.
N

100 METERS

The ball of fire. (Courtesy of LANL and the National Atomic Museum.)

All photos courtesy of the Los Alamos National Laboratory unless otherwise noted.

Tennis in front of the Big House when Los Alamos was a boys' school. The wooden object at the right is a toboggan slide. (Courtesy of the Los Alamos Historical Museum Photo Archives.)

Facing page: The road to Los Alamos, 1944. Notice the rocks used for the guardrail.

Building B in the Tech Area. (Courtesy of LANL and the Los Alamos Historical Museum Photo Archives.)

Dorothy McKibbin in 1983.
(Photo by the author.)

Facing page: Dorothy Jenson getting off the train at Lamy, N.M., 1943. "Our instructions were to proceed by bus to 109 E. Palace in Santa Fe. No talking to anyone." (Courtesy of the Los Alamos Historical Museum Photo Archives.)

Above, left: Jack M. Hubbard, head meteorologist for the Trinity Site test, June 1941.

Above, right: John von Neumann. (Courtesy of LANL and the National Atomic Museum.)

Left: Hans A. Bethe.

J. Robert Oppenheimer.

Above, left: Gen. Leslie R. Groves.

Above, right: Kenneth T. Bainbridge, director of all on-site activities at Trinity.

Left: Enrico Fermi. (Courtesy of LANL and the National Atomic Museum.)

Ernest O. Lawrence, Enrico Fermi, and I. I. Rabi.

Edward Teller (center) and Norris Bradbury (right) at a party in the lodge.

The meeting of two worlds: Enrico Fermi is introduced to Maria Martinez the famous potter of San Ildefonso pueblo (probably at the December 1945 dance). (Courtesy of the Los Alamos Historical Museum Photo Archives.)

Chianti bottle used to celebrate the first controlled nuclear reaction in Chicago, December 2, 1942. (Courtesy of the Argonne National Laboratoi and the National Atomic Museum.)

CHAPTER THREE

Theoretical Considerations

All things considered, the actual construction of Trinity Site proved relatively uneventful. The theoretical considerations of what might happen to the area because of the detonation, however, proved quite another matter. In examining these issues, the scientists raised three questions which have still not been fully settled.

The most important of these—and this may be an understatement—was that a nuclear explosion might somehow ignite the atmosphere and thus destroy all life on earth.

By 1945, the idea that the world might come to a violent and abrupt end was hardly new. Its chief proponents hitherto, however, had been the millennial Christians. Reading the books of Daniel and Revelation literally, these people predicted that Jesus would return bodily to inaugurate the promised thousand years of peace. Then time would come to an end. Adventist William Miller of upstate New York gathered many followers for these views in the early 1840s. Twice he actually set dates for the process to begin. Although these ideas faded somewhat with the Civil War, they were revived during the late nineteenth and early twentieth centuries. Numerous conferences, speakers, and publications moved the millennial concept into many of the mainline Christian denominations. The advent of World War I seemed to lend credence to this position. Many felt that the world had entered the "last days."[1]

With the advent of subatomic physics in the new century,

however, the question became recast. In the new version, the world would be brought to an end, not by the predictions of Daniel and Revelation but by the auspices of science. It would be inaugurated not by the return of Jesus, but by a nuclear error of man.

The early physicists were not unaware of this potential for disaster. In 1903, Ernest Rutherford made the "playful suggestion" that "could a proper detonator be found, it was just conceivable that a wave of atomic disintegration might be started through matter, which would indeed make this old world vanish in smoke."[2] In 1920, physicist A. S. Eddington suggested that the source of the sun's energy might be the conversion of hydrogen into helium. Two years later Nobel Prize winner Francis William Aston spoke to a Philadelphia audience on the same theme. Warning against "tinkering with angry atoms," Aston noted that if the hydrogen in the atmosphere were somehow transformed into helium, "this most successful experiment might be published to the rest of the universe in the form of a new star. . . ."[3] In the 1930s, Leo Szilard and George Gamow began serious analysis of whether a nuclear chain reaction could accidentally set off an endless sequence. What would later become a staple of science fiction writers was common knowledge among the world's physicists by the early twentieth century.

The principle involved here was not nuclear *fission* or the splitting of the heavy atoms at the upper end of the periodic table—uranium or plutonium. Rather, it involved nuclear *fusion*, or the forcing together, under great heat and pressure, of those lighter atoms at the lower end of the table—hydrogen, helium, and nitrogen. This process was believed to be that by which the sun and stars shone. It required enormous temperature and pressure to fuse these atoms, however, and these conditions existed nowhere on the face of the earth.[4] But would they somehow be created in the interior of a nuclear explosion? This was the question.

In July of 1942, a group of physicists began discussion of this issue in LeConte Hall on the campus of the University of California. Edward Teller made some of the initial calculations, and his figures indicated that the bomb would, indeed, create enough heat to ignite the earth's atmosphere.[5] Oppenheimer immediately called a halt to the meetings and sought out Arthur H.

Compton, head of the Metalurgical Laboratory of the University of Chicago, who was his immediate superior.

Compton was vacationing with his family at their summer cottage in Michigan when he received Oppenheimer's urgent call. When Oppenheimer finally arrived by train from California, the two men had long discussions by Oswego Lake on this matter. It proved a morning both would long remember. Oppenheimer then returned to Berkeley, and the scientists met frequently during the next three months to continue to analyze the issue. In a 1959 interview with author Pearl Buck, Compton recalled that he took the initiative in the early days of this discussion. If the calculations showed that the chances of igniting the atmosphere were more than approximately three in one million, he said, he would not proceed with the Manhattan Project. Nothing, not even the horrifying possibility of a Nazi victory, was worth this. Revised calculations showed figures slightly under that, and so the project continued.[6]

As the scientists reexamined this issue, they discovered that Teller's early figures had been in error. He had neglected to include the loss of heat to the cooler air which would surround any nuclear explosion. It took several years of intensive calculations, however, to determine the correct figures. In the interim things progressed as usual.

From 1944 on, as the bomb became more and more of a certainty, this question of possible ignition of the atmosphere was examined again and again. All the best physicists in the nation were involved in it: Oppenheimer, Fermi, Szilard, Bethe, Teller, Wigner, Compton, and others. At Los Alamos, Hans Bethe, head of the Theoretical Division, played the crucial role. His calculations showed that the extreme pressures and temperatures reached in the interior of the Trinity explosion would not be high enough to fuse the hydrogen with either nitrogen or helium to form a new star. "We all put our faith in Bethe," recalled Bainbridge.[7]

In an interview thirty-seven years later, Bethe disclaimed complete responsibility. "It was not just my theories," he said. Edward Teller's group at Los Alamos also seriously examined the question.[8] Teller's revised figures agreed with Bethe's. There was no danger. Physical chemist John Magee recalled that shortly before the test, Teller gave a talk at one of their weekly meetings, assuring everyone that a vast safety factor existed. At the site itself, few people took the question seriously. They were not being frivolous; they simply did not think there would be any problem. "I wasn't ever worried about it," Magee said. "I was convinced by what Edward said."[9]

In 1975, the government declassified the December 2, 1943, report by Teller and Emil J. Konopinsky that declared there was "a safety factor of about 60 preventing ignition of the atmosphere by the starting temperature of 3 MV [million electron volts]." They concluded that the turning of the earth into a star was "impossible"—even with an enormous bomb loaded with 1,000 cubic meters of fuel (the Trinity explosion had about twelve pounds of plutonium).[10] In 1975, Eugene P. Wigner of Princeton also recalled his role in these discussions. "I made the calculations," he noted, but he found no evidence that a runaway reaction was possible—"perhaps to my regret," he added, meaning that if one had been possible, no atomic weapons would ever have been built.[11]

After the war ended, this question of the possible ignition of the atmosphere became public property. The scientists repeatedly assured the nation that the danger had been overrated.[12] Man need not worry about disintegration of the earth, said Ernest O. Lawrence, because "we're winning greater, not less, control over nature."[13] Hans Bethe was especially vocal on this issue. He decried the loose talk about it and said that it came from scientists who were totally ignorant of the problem.[14] In 1946, he addressed a special Senate hearing on atomic energy and told them that the ignition of the atmosphere or the ocean would be possible only if one took an enormous amount of water or air— perhaps the size of the sun itself—and then raised this mass, all at the same time, to a temperature of many millions of degrees. Only then could there be a self-sustaining nuclear reaction. "I think I should say," he assured the senators, "that before we made our first test of the atomic bomb, we were sure on theoretical grounds that we would not set the atmosphere on fire."[15]

With each advance in nuclear armaments, the issue of possible ignition of the atmosphere has been raised again. Just before the first hydrogen bomb was detonated on November 1, 1952, Edward Teller asked Gregory Breit of Yale to reassess the situation. Breit's notes revealed Teller's fear that "the consideration of the wartime period might have been too hurried," and that the dangers of a "runaway super [H-bomb] have increased the importance of the question." The question was also seriously discussed in the early 1950s at Oak Ridge.[16]

In November 1975, the distinguished *Bulletin of the Atomic Scientists* raised the issue yet again. Radiation physicist Horace C. Dudley maintained that under certain circumstances—deep under the ocean or far under ground, for example—a nuclear reaction might find conditions favorable enough to become self-sustaining. Bethe and other physicists, however, dismissed Dudley's theories as nonsense.

The most recent argument along these lines may be found in Jonathan Schell's *The Fate of the Earth* (1982) and in the reports of the Washington, D.C. conference on the Biological Consequences of Nuclear War, which was held in the fall of 1983. *New Yorker* writer Schell argues powerfully. He suggests that if enough nuclear explosions occur, the earth's atmosphere will be so damaged that "We—the human race—shall cease to be." Although the conference scientists utilized more prosaic language, they reached identical conclusions. They forecast that a full-scale atomic exchange would result in a "nuclear winter" that would encircle the globe. One scientist suggested that this would lead to "the extermination of *Homo sapiens.*"[17]

This question will be raised, as well it should, with every turn of the nuclear screw. But it received its first concerted discussion in the months preceding the July 16, 1945, Trinity test. Those atomic scientists analyzed what had to be the most agonizing question in all of human history.

While the theoreticians may have concluded that no danger existed, that did not preclude on-site banter. On the night preceding the test, Enrico Fermi began taking wagers as to whether the bomb would ignite the atmosphere, and if so, whether it would destroy just New Mexico or the entire world. He guessed that if the nitrogen in the air were ignited, it would go only about thirty-five miles or so. Many of the scientists chose to watch the test from a considerable distance away. When they saw the sky light up so brightly, several worried momentarily that they had miscalculated and that all the experimentors at the three shelters had been incinerated.[18]

Fermi may have been trying only to break the tension through black humor, but both General Groves and Kenneth Bainbridge were furious when they heard about his comments. They feared that such loose talk might paralyze some of the enlisted men with fright. Indeed, when the bomb exploded, the shocked reactions of several people showed that they had little or no comprehension of what was going on. Enlisted man Frank Diluzzo's first thought was: "My God, we've ignited the atmosphere." As photographic specialist Berlyn Brixner watched the ball of fire grow, he wondered, "Will it ever stop?"[19] Even the military was edgy. As the fireball rose in the sky, one officer gasped, "The longhairs let it get away from them."[20]

Although considerably less dramatic than the question of atmospheric ignition, the scientists seriously considered a second theoretical question, which also had considerable social consequences: How would the gigantic blast affect the nearby area? Indeed, the documents available suggest that as much effort was devoted to the question of air blast and accompanying ground shock as to any other possible aftereffect.

The bomb, of course, was designed as a military weapon. From a military point of view, the blast wave would be the most effective agent to damage enemy installations. Consequently, they wanted elaborate measurements of its power. Moreover, blasts of high explosives formed a considerable part of the Los Alamos experience. Explosions often disturbed the silence of the mesa area, but local residents would shrug them off as "thunder." "An experimental blast in a nearby canyon would rock the school to its expensive foundations," recalled Ruth Marshak, who taught English there. "There would be a frightened silence in the classroom, and then we would begin again on Henry Wadsworth Longfellow."[21]

With this in mind, the scientists studied the local effects from all the major historical explosions they could find. For example, on December 6, 1917 in Halifax, Nova Scotia, a French munitions ship exploded in the harbor killing 1,600 and leaving 10,000 homeless—perhaps the largest single accidental explosion in history. On September 21, 1921, an explosion of a pile of ammonium nitrate at Oppau, Germany created a crater 130 yards wide and 45 feet deep. Shocks from this blast were felt 50 miles away.[22] Closer to home, a bomb-filled boxcar at tiny Tolar, New Mexico, exploded on November 30, 1944, killing one man and leaving a

75-foot crater ten feet deep. Several Los Alamos scientists even visited the famous Meteor Crater, a gigantic impression in the earth about 30 miles east of Flagstaff, Arizona. As they viewed the mammoth crater, they asked themselves if this might not be the model for what the Trinity explosion could do.

The scientists worried about possible damage to nearby civilian areas from the blast, or even a possible earthquake. Because of security precautions, they desperately hoped to avoid major lawsuits. Groves especially worried about this. As a result, in mid-April 1945, Oppenheimer called in the best expert in the nation on such questions, Harvard seismologist L. Don Leet, then working with the Underwater Sound Laboratory in Massachusetts.[23]

Leet poured over the carefully procured geologic maps of the Jornada with a fine-toothed comb. He examined the local fault lines and worried about the potential collapse of some old, abandoned coal mines near Carthage. He also studied the blast statistics from the May 7 100-ton shot. Finally, he concluded that there was no possibility of property damage within 5,000 yards because of earth shock:

> As the distance increased beyond 500 yards, the hazard of damage from ground vibrations decreases very rapidly until even at a few thousand yards and certainly at tens of miles it ranges from fantastically small to utterly impossible. The slamming of a door, or someone walking across the floor of a house in Socorro, Carrizozo, or other surrounding towns will produce many times more displacement in floor, wall, and ceiling panels of the house than will any grand vibrations from a 5,000-ton shot, or two 50,000-ton shots.[24]

Leet warned, however, that there might be potential damage from the airborne concussion. Ordinary civilians seldom distinguished between the two, and people in Tularosa, forty-five miles away, had reported feeling an "earthquake" after the 100-ton shot. Leet urged that strong-motion seismographs be stationed at numerous places in the area: San Antonio, Tularosa, Carrizozo, and Base Camp. He also suggested that the instruments be put inside the buildings to show the relationship between ground and air waves. "This will be very useful," he observed, "if complaints are received."[25]

When the bomb was detonated, numerous people felt the air-burst from it. Windows rattled, and some were actually broken as far away as Silver City and Gallup. Although the air wave was most marked in the direction of San Antonio and Socorro, even there it was well below the magnitude of serious damage. Over fifty miles away, Leet himself found the air wave from the blast barely perceptible. "It was like a greatly attenuated thud from the firing of a distant gun," he noted.[26]

From a purely scientific point of view, the Trinity blast created two new elastic waves that had never before been recorded by seismographs. Leet termed these waves extremely important and in a later article noted that Trinity provided "one of the most important records of earth motion in the history of seismology."[27] All the scientists' planning regarding earth shock and air blast proved accurate. No civilian property was damaged. No earth-quake occurred. The blast was such that many local residents slept right through it.[28]

Related to the question of potential earthquake was that of a possible firestorm. When the 100 tons of TNT were exploded, the surrounding area was scraped by a bulldozer to prevent any range fire. Some scientists expressed concern that the nuclear explosion might set the entire Jornada ablaze. Here, however, the sparse vegetation of the valley proved beneficial. Unlike the western California deserts, where the grasses are thick enough to support firestorms, range fires on the Jornada quickly extinguish themselves. The vegetation grows too far apart to keep them alive.

The last item of major social concern the scientists considered, and the one that turned out to have the most significant long-term consequences, involved the question of radioactive fallout. Although the fallout issue eventually proved more important to the nation than atmospheric ignition or the blast wave, discussion of it did not loom large on the scientists' initial list of priorities. Fallout, as yet, was not a part of the human experience. Moreover, in the early planning stages for Trinity—indeed, right up until shortly before the test itself—Jumbo was seen as a probable safeguard for any "fizzle." Its presence probably also inhibited consideration of other fallout-related questions.

Although both the 1941 British MAUD report and an October 13, 1944, document by George Kistiakowsky mentioned the fallout question, the issue was first seriously raised at Los Alamos by physical chemists Joseph Hirschfelder and John Magee. In late 1944 they had finished one assignment and happened to be available when Hans Bethe needed someone to consider all the possible blast effects from Trinity. One morning in early April of 1945, Hirschfelder came dashing into Magee's office and shouted, "What about the radioactivity?" Immediately the scientists realized that the fallout question would have significant medical and biological consequences.[29]

Hirschfelder and Magee then began to devote all their time to this issue. They combed through such books as *Design of Industrial Exhaust Systems, The Physics of Blown Sand,* and *Micrometrics,* a story of air pollution from industrial smokestacks. They also began numerous experiments on samples of sand and dirt brought from under Ground Zero. They put the soil under immense pressure, incinerated it with blow torches, dropped it from various heights, and blew it with fans. They did this to try to arrive at an average particle size and probable distribution ratio of the fission products. In addition, they pored over the maps of the area to see how the wind might distribute the materials.[30] It was because of their recommendations that the area immediately under the bomb was paved over—to reduce the size of particle pickup.

The early predictions of the amount of fallout by Joseph Hirschfelder and John Magee were decidedly pessimistic. Drawing on the conclusions that Herbert L. Anderson derived from the May 100-ton shot, the two chemists postulated on June 16 what might happen in the full-scale test.[31] They assumed a cloud that would rise to only about 12,000 feet. If the active material were to condense on the surface of the dust particles at the same ratio as in the 100-ton shot, they predicted that a large and dangerous amount of radioactivity would soon sediment onto the nearby towns. They estimated that 49 percent of the activity in the cloud would fall within the first two hours and twenty-two minutes.[32] These estimated figures, had they been correct, were so lethal as to make the test virtually impossible. Copies of their memos to this effect all reached Oppenheimer's desk. It was not until July 6, 1945, that Hirschfelder and Magee recalculated their fallout predictions, lowering their estimate by a factor of 2 to 10.[33]

In early May, Louis Hempelmann inserted 1,000 curies of fission products from Hanford into the 100 tons of TNT on the wooden tower. They hoped that the distribution of radioactivity from this trial run would provide clues for the real thing in July. The blast—the world's largest controlled explosion up to that date—sent a cloud of smoke and debris up to between 12,000 and 14,000 feet. It was held at that level because of a slight inversion. Then the cloud drifted east. The explosion also created a shallow crater, about twenty-nine feet wide and over four feet deep. When the scientists monitored the region, they found measurable activity only within a radius of thirty feet from the tower. Even at the center of the crater, the readings were very low. Since the powdery dust showed slight radioactivity, the medical group recommended the use of protective booties and respirators for all those who entered the crater.[34]

The results of the test shot were carefully calibrated, but they were not terribly helpful. The blast yield from 100 tons of TNT could be accurately estimated, but both the yield and wind pattern from the July test still remained uncertain. It "was obvious to everyone," Sam K. Allison wrote to Oppenheimer on the 100-ton shot, "that answers to questions of greatest interest were in general not obtainable. . . ."[35]

The Los Alamos physicians were another group especially interested in questions of fallout raised by the May 7 test. Colonel Stafford Warren, medical director of the Manhattan Project, followed the smoke cloud from the 100-ton shot for several hours. Much to his dismay, he discovered that a brisk 30 MPH wind carried it right over the towns of Carrizozo and Roswell. His fellow physicians James Nolan, chief medical officer at Los Alamos, and Louis Hempelmann joined him in this concern. They began to realize that the wind pattern on the day of the test would be of great importance. Warren wrote Groves of his concern.

The physicians always felt that they received relatively little support from their Los Alamos colleagues on the fallout issue. Most of the men in charge of the test had been trained as physicists. The doctors felt that the physicists focused their attention primarily on those events leading up to the moment of the blast. If everything went as planned, their task was completed. What happened afterwards, however, became the responsibility of the physicians. They were responsible for the radiological safety of the populace, before the field had really been defined.

On June 20, 1945, James Nolan flew to Oak Ridge to speak to General Groves about potential fallout dangers. He carried with him a report written by McGee, Hoffman, and Hempelmann. It assumed that the cloud would rise to about 12,000 feet. Nolan waited in Groves's outer office while the general read the document. When Groves called Nolan into his office, he was not pleased. "What are you," he said to the physician, "some kind of Hearst propagandist?"[36]

While Nolan and Hempelmann seemed to swing little weight with Groves, the general always placed implicit trust in Stafford Warren, who also served as his personal physician. Only weeks before the scheduled test, Warren flew to Washington with a more detailed map of the probable fallout pattern. The area of immediate concern involved a radius of thirty miles from Ground Zero. "Well," Groves told Warren, "you sure play hob with everything."[37] It was only then that the army began laying serious plans for potential evacuation of the surrounding area.

The problems of estimating the Trinity fallout tied directly into two factors that would not be known until the instant of detonation: the yield of the explosion and the winds of the day. There was no agreement among the scientists as to how powerful the blast would be. Many felt that the bomb might fly apart on its own, before the mass of plutonium was split in the chain reaction. In a pessimistic moment, John Magee asked Hans Bethe why they could not be realistic and calculate the effects for a bomb equivalent to 1,000 tons of TNT (a rather low yield). Bethe, however, said no. He instructed them to continue figuring on a much larger blast.[38]

Magee was not alone in his relative pessimism. William Laurence recalled that many scientists expressed great uncertainty up to the very last minute. Even General Groves later confessed that, in spite of his outwardly optimistic demeanor, he estimated only a 40-60 chance of success.[39]

This uncertainty can be seen in the betting pool—one dollar a chance, limited to one per person—in which over a hundred scientists participated. Oppenheimer guessed that the explosion would be equivalent to 300 tons of TNT. This was a very low estimate, not much more than thirty ten-ton blockbuster bombs—

in other words, a "fizzle." Stan Ulam and Hans Bethe guessed in the 5,000–7,000-ton range.[40] Except for Edward Teller, who took 45,000 tons, most of the others took similarly low figures. The winner was I. I. Rabi, who arrived late and found all the lower numbers taken. Trying to be polite, Rabi guessed 18,000 tons—considerably higher than any of the others. The explosion was eventually rated at close to 20,000 tons, and Rabi walked away with a large wad of bills.[41]

In addition to the question of yield, the fallout distribution also depended on the weather. Winds move in different directions at different levels of the atmosphere, of course, and no one could say to which level the blast might penetrate. The scientists estimated that a low yield would cause the radioactive cloud to move to the northwest. A moderate yield might divide the blast. A high yield would send the cloud to the northeast. In fact, the cloud was split into three parts—something no one had predicted. So, the weather conditions at Trinity began to loom as one of the most important factors.

CHAPTER FOUR

The Question of Weather

Weather has been termed "the most violent variable in man's plans and enterprises."[1] Certainly this was true for the Trinity explosion, for a fierce summer thunderstorm came in unexpectedly, and played havoc with all the scientists' plans. Here one must turn to the story of the chief meteorologist for Trinity Site, Jack M. Hubbard. Hubbard's journal, faithfully recorded at the end of each day, has just been released to the public. It is one of the most intriguing Manhattan Project documents to emerge within the last twenty years.[2]

Born June 27, 1916, Jack Hubbard grew up in the Pacific Northwest, receiving his B.S. at Washington State University in Pullman and his M.S. at the University of Washington in Seattle. After studying meteorology at the California Institute of Technology (Caltech), he worked for over a year as a meteorologist for Pan American–Grace Airways in Buenos Aires. There he became acquainted with unstable, tropical air masses. Afterwards, he also worked as a civilian scientist attached to the Air Force Research Center at Caltech and with Northwest Airlines to draw up a detailed meteorological study of all the Asiatic coastal cities.

In March of 1945, Hubbard had just completed this assignment when a letter arrived asking him to join the Manhattan Project. Project officials had been negotiating with another meteorologist, but their discussions had fallen through. They needed a long-term forecaster immediately. In early April Hubbard accepted the assignment.

Meteorologists did not have much weather data to work with, for there were few records for the Jornada del Muerto area. A comprehensive weather summary for the entire nation (available at Caltech), the Airways Meteorological Atlas for the United States, and local data from the White Sands National Monument were all that existed. Because of the influence of the Oscura Mountains, they were only moderately helpful. The Oscuras caused local variations in both wind velocity and direction. Conditions at Trinity were different from the weather stations at nearby Albuquerque, Roswell, and El Paso.[3]

After arriving in Los Alamos, Hubbard worked closely with Kenneth Bainbridge. Bainbridge stressed the importance of the weather for the test and outlined the probable problems with atmospheric fallout. He also put Hubbard in charge of a team of enlisted men, all of whom had had previous meteorological experience.

Hubbard followed the standard techniques of long- and short-range forecasting developed by Caltech and the AAF Weather Service. Operating from a sixteen-by-sixteen-foot portable hutment at Ground Zero, he utilized all the latest equipment, including, among others, two portable AAF weather stations, two field radar sets, a radiosonde (which gave humidity and temperature readings at various altitudes), and some pilot balloon wind observing sets. He studied the weather maps from Kirtland and Alamogordo air bases and also worked closely with both Caltech and the AAF Weather Service in Washington, D.C. to obtain state data for the preceding eighteen months.

Hubbard's first assignment was to forecast the weather for the 100-ton test in May. Moreover, he was allowed to choose the day for the May shot. Accordingly, on April 18 he sent out inquiry letters to all group leaders concerned asking them what weather conditions they would prefer for the trial run. From their replies he compiled a comprehensive ideal weather picture: visibility over forty-five miles, humidity under 15 percent, clear skies, surface winds calm, westerly winds aloft, with a preferred Bernoulli effect over the Oscura Mountains. Temperature lapse rates aloft should be no more than slightly stable, to prevent subsidence of the cloud back to earth, ground inversion about 1,500 feet, and no precipitation in the area within twelve hours of the operation.

On April 23, Hubbard made his first long-range forecast and concluded that these conditions would occur around the first week of May. On May 2, he confirmed that the original forecast was operating properly and fixed the final date as the morning of May 7. During the evening of May 6, a low-pressure trough came through the camp and brought with it heavy wind and blowing dust. Stafford Warren, who had been caught in the storm, confronted Hubbard and asked how it would be possible to hold the test the next day. Hubbard replied that the trough passage was just what he needed to bring the winds around to the right direction in about six to twelve hours. At dawn the next day, when the 100 tons of TNT were exploded, the winds were exactly as predicted. In his official report on the 100-ton test, Hubbard proudly noted that between April 23 and May 10, only two days filled all the required weather specifications. He called one of them perfectly.[4]

This initial success probably convinced Hubbard that he could repeat his performance for the real thing in July. But in the Southwest, July offers much more turbulent weather than does May. Even more important, whereas Hubbard was permitted to pick the ideal date for the May 7 test shot, he was not allowed to do this in July.

International politics suddenly began to play a deciding role in the Trinity operation. Groves's instructions to Los Alamos were to prepare an atomic weapon "as soon as possible," and pressure from the military increased dramatically during late June. Groves and Oppenheimer had several loud arguments over the issue of scheduling. Germany had officially surrendered on May 8, and Harry Truman was to leave for an international conference with Joseph Stalin and Winston Churchill in Potsdam, a suburb of Berlin, on July 6. The conference had originally been scheduled earlier, but Truman had asked that it be postponed until July 15. This was to give the Los Alamos people more time.[5] The key issue of discussion obviously would be how to end the war in the Far East. If Truman had knowledge of a successful test at Trinity, he would have a powerful negotiating weapon to use at the conference. On July 4, Bainbridge, who was on the telephone to Washington, asked Hubbard what he thought of the day of July 16 as a test date. Hubbard replied negatively.[6]

Meanwhile, as he did with the May 7 test, Hubbard had requested the various group leaders to send him what they deemed desirable and undesirable weather conditions for the real thing. From these, he compiled the following list:

Best conditions for the operation.
  A. Visibility greater than forty-five miles.
  B. Humidity below 85 percent at all levels.
  C. Clear skies.
  D. Temperature lapse rate aloft slightly stable to prevent dropping of the cloud.
  E. Little or no inversion between 5,000 and 25,000 feet to allow cloud to reach maximum altitude.
  F. A thick surface inversion or none at all to prevent internal reflections and mirage effects.
  G. Winds aloft fairly light, preferred direction from between 6° south of west and 25° south of west. Steady movement desirable to anticipate track of cloud. Horizontal and vertical wind shears desirable for maximum dissipation of the cloud, although such a condition increased the tracking problem.
  H. Low-level winds light and preferred drift away from Base Camp and shelters.
  I. No precipitation in the area (thirty-five miles) within twelve hours of the operation.
  J. Predawn operation desired by the photographic group, although 0930 operation considered best for thermals dissipating the lower levels of the cloud.

Conditions least favorable to the operation.
  A. Haze, dust, mirage effects, precipitation, restrictions of visibility below forty-five miles.
  B. Humidity greater than 85 percent at the surface or aloft, which might result in condensation by the shock wave.
  C. Thunderstorms within thirty-five miles at the time of operation or for twelve hours following.
  D. Rain at the location within twelve hours of the operation.
  E. Surface winds greater than 15 MPH during and after the operation.
  F. Winds aloft blowing toward Base Camp or any population center within ninety miles of the site.[7]

On June 21, Hubbard made a long-range forecast for the July test. He predicted that the desired conditions would probably occur July 13, 19, or 20. Nine days later he revised this slightly. July 18–21, inclusive, were his first choices; July 12–14 were his second.[8] The middle of July found Trinity resting under oscillating air masses: cool, stable ones from the north alternated with moist, warmer ones from the south. Hubbard's first choices were when Trinity lay under the former.[9]

During the first week of July, however, the date of July 16 became fixed "from above." The person making the decision was never mentioned in any of the documents, but it was General Groves. "By fixing the date of operation for the morning of the 16th, weather permitting," Hubbard noted, "it became necessary to tailor the operation and adjust certain aspects of the operation to the expected wind structure."[10]

So, instead of tailoring the operation around the desired weather conditions, as he did for the May test shot, Hubbard faced a *fait accompli.* "This changed the weather-operation relationship," he wrote, "and introduced a consideration of the minimum specifications under which the operation could be conducted."[11] Although couched in jargon, it is clear in his final report on Trinity that Hubbard considered this a blunder of the first magnitude.[12]

With these new conditions, pressures on the weather crew began to mount. Beginning with June 25, hourly observations were kept by the weather stations at Base Camp and Ground Zero. As predicted, the tropical gulf air mass arrived on July 10, filling the air with a slight haze because of its salt particles. It dominated the Trinity area for six days and gave the meteorological team a good chance to probe it thoroughly.

Stafford Warren recalled that often the men at Trinity would argue past midnight, wondering what might happen and where the radioactive cloud might go. Several observers were skeptical that the weather could be scheduled with such precision as to avoid the nearby settled areas. "Anybody knows," General Thomas Farrell observed in one discussion, "that you can't predict the weather that well."[13]

During the second week of July, Frank Oppenheimer, Robert's younger brother, and a physicist in his own right, and J. F. Mattingly set up about a hundred smoke pots on the Sierra Oscura range. These were lit on July 7 and again on July 14–15. The purpose was to see how the atmosphere behaved at dawn in close proximity to the escarpment. Hubbard flew in an L-45 Beechcraft over the escarpment to try to estimate how the radioactive cloud would behave. He concluded that it would probably settle in the valleys on both sides of the range and that a gradual downslope would also carry the cloud from Ground Zero over to the shelter at South 10,000. The hope was that the dilution would be such that Carrizozo and other nearby towns would have no problems with radioactivity.[14]

Other weather experts were also called in. Colonel Ben Holzman, author of the standard pilots' weather manual and one of the men who had selected the date for the Normandy landing, was flown out to offer his suggestions. Hubbard also had daily phone conversations with Major John Wallace, who was stationed at the Pentagon. On July 12, Wallace pulled from the punch card file in Washington the date of July 28, 1906 as the analog date (the day having the most similar weather in the last forty-five years) to that of July 16. University of Washington Professor Philip E. Church, an expert in the dilution of radioactive fallout with experience at Hanford, also arrived, although Hubbard took little stock of his opinions. In addition, Hubbard received the daily special forecasts from the AAF Weather Division in Washington and the regular ten-day forecast via airmail.

On July 6, Hubbard made another forecast for July 16. He concluded that the predawn would find Trinity in a period of stagnation. The area would be dominated by sluggish, northward-moving gulf air, and there would be easterly winds and afternoon thundershowers. From the standpoint of forecasting, these are difficult conditions.

Bainbridge, who was in frequent contact with Washington, tried for an earlier test date, but on July 12 he received word that the date had been definitely fixed for the morning of July 16. From then on, everyone in the weather group knew that the test would have to be carried out under conditions that would be less than ideal. Hubbard's diary recorded his reaction. "[R]ight in the middle of a period of thunderstorms," he wrote, "what son-of-a-bitch could have done this?"[15] They were told that the decision had something to do with a conference to be held in Potsdam. Hubbard immediately asked Bainbridge if they could wait until July 18–21, when the tropical air mass would have cleared the Trinity area. Bainbridge, however, said that July 16 had become absolutely firm, and that he had had no say in the decision.

At 11:00 A.M. on July 15, Hubbard issued another forecast where he predicted that the next day would see the winds light and variable, from east to west below 14,000 feet, and west southwest above 15,000 feet. What he did not predict were the fierce local thunderstorms which moved into the area around 2:00 A.M. that night, carrying gusts of 30 MPH winds with them.[16]

It is a meteorological truism that at any moment, there will be between 3,000 and 4,000 thunderstorms somewhere on the globe. This one over Trinity, however, appeared at just the wrong time. Moreover, it was especially severe. In 1982, both Sam P. Davalos and Jerry Jercinovic confessed that they had not yet seen a storm that equaled it in intensity.[17] Although the storm was localized—little rain fell at Ground Zero itself—it drenched the S-10,000 shelter and Base Camp, where most of the personnel were stationed.

No dramatist could have staged the setting for the Trinity atomic detonation with more power than it had in fact. The tension could be felt everywhere. Cloud-to-cloud lightning lit up the tower at intervals, and the rain fell in torrents. The ground was strewn with puddles, and numerous scientists worried about possible electrical shorts. A roster of distinguished visitors anxiously waited about twenty miles away. Bainbridge, Oppenheimer, and Hubbard each had the power to veto the shot, for any reason whatever. One vote alone would suffice. Because of the weather, many people urged Oppenheimer to cancel. The men were so exhausted, however, that any postponement would have required a couple of weeks for recuperation.[18] Neither Bainbridge nor Hubbard, for example, had slept in over two days. There was no possibility that the test could have been rescheduled for, say, the next day, or even the day after.

I. I. Rabi, who had been brought in from MIT as an observer, killed the time by playing poker. He observed later that in spite of all their careful planning, the real decision on detonation did not rest with the Los Alamos scientists. Instead, everything lay in the hands of the meteorologists.[19]

Opinion was mixed as to whether they should proceed. Numerous people at Base Camp became instant weather experts and offered Oppenheimer their advice. Some, such as Robert

Bacher, pressed for firing, but the majority favored postponement.[20] Monitors Joseph Hirschfelder and John Magee sat waiting in a car, listening to the rain on the roof. They can't have
the test, Magee said. "The hell they can't," replied Hirschfelder.
"They've brought in all these high officials, and they can't delay.
It's going."[21]

There was probably as much pressure on Hubbard as on anyone. Everything he did not want was present: rain, high humidity,
inversion levels, and unstable winds. Rain was undesirable because it might "scrub" the cloud and bring down a high concentration of radioactive particles in a small area. Safety lay in
spreading the material as widely as possible. The only item on
which everyone agreed was that the bomb should not be exploded in the rain.

Unstable conditions and humidity over 80 percent offered another danger because of fear that the blast wave might somehow
induce a thunderstorm. Both traditional weather folklore and
the infant science of pluviculture suggested that loud explosions
could actually bring on rainfall. (Why does it rain so often on
the Fourth of July? was a standard question.) Working with Hubbard, Joseph Hirschfelder had explored this question in detail.
They both concluded that, given proper conditions, the blast
might, indeed, induce rain.[22]

Hubbard's diary reveals the pressure. After midnight on July
15, his airborne instruments showed that the gulf air over Trinity
had deepened to a level of 17,000 feet and that it contained two
inversion levels. This, too, was potentially dangerous, for if a
rising column of gas meets decreasing pressure, it can expand
and undergo an adiabatic decrease in temperature. If it meets a
strong inversion, it will be colder and more dense than its sur-
roundings, and as soon as its upward momentum is expended,
it will then fall rapidly back to earth. This process is known as
"looping" and can sometimes be seen on cold mornings around
industrial smokestacks. Hubbard feared that if the explosion
produced only a moderate yield, it would take the radioactive
cloud up and then bring it right back down again. This would
seriously contaminate everyone in the immediate area.[23]

The winds proved even more worrisome. The men in charge
of monitoring the expected fallout prepared for two alterntives—
a "North Blow" or a "South Blow," depending on the direction

of the wind. But the winds actually shifted a full 360 degrees in the twelve hours preceding the test.[24] Had there been an instrument malfunction or an error on Hubbard's part, Socorro, thirty miles to the west or Carrizozo and Roswell, thirty and ninety-five miles to the northeast and east, respectively, might have been dusted heavily with radioactive fallout. Hubbard needed a wind from the south-southwest, operating within carefully specified degrees, to carry the fallout over uninhabited areas and miss these nearby hamlets. He called it correctly. That such errors are possible can be seen in the cases of both a 1954 H-bomb test in the South Pacific and a 1958 atomic explosion at the Nevada test site. In the former, the cloud broke through unexpectedly to a higher wind level and dumped radioactive materials over several islands that had been presumed safe. In the latter, the cloud was carried 220 miles southwest to the Los Angeles area by strong, unpredictable winds. There it became trapped under an inversion level for several days before it dispersed. Normally a man of great self-assurance, Hubbard was decidedly uncomfortable for several hours that night. "He was really on the spot, no doubt about that," one scientist recalled.[25]

Around noon on July 15, Generals Leslie Groves and Thomas Farrell arrived at Trinity. At 4:00 P.M., a conference was held at the McDonald ranch house with the two generals, Oppenheimer, Bainbridge, Colonel Ben Holzman, Richard C. Tolman of the NDRC, and Hubbard. The major item of discussion was the weather. Hubbard maintained that he had been against the July 16 date ever since it had first been mentioned. But, he said, the test could probably be safely conducted "in less than optimum conditions, which would require sacrifices."[26]

After some discussion, Groves and Oppenheimer retired to another room. When they returned, they decided to postpone any final decision until after the last, 2:00 A.M. weather conference. Then they would see how extensive the sacrifices would have to be. The meeting then broke up, and each went to his respective job.

Around 10:00 P.M., the weather instruments indicated that the period of stagnation had ended. Trinity was about to undergo a sequence of trough passage and violent weather. From 10:00 P.M. to midnight, Hubbard charted both the instruments and the sky. Everything seemed awry. At 11:00 P.M., the winds were such that, had they held in this position, the shelter at N-10,000 would have been inundated with radioactive debris. At 1:15, after the passage of the trough, the winds had shifted considerably, but with little effect. In this position, they would have inundated S-10,000 and made any escape route through S-10's nearby Mockingbird Gap impossible. Several hours were needed for the winds to right themselves for the prescribed conditions.

The initial time for the shot was 4:00, but that hinged on the final clearance from the 2:00 A.M. weather conference, to be held at Base Camp. Owing to last-minute checking, Hubbard arrived there eight minutes late, and he was met outside by a nervous Oppenheimer. Hubbard told the director that things did not look good for the 4:00 A.M. test but that all should be acceptable between 5:00 and 6:00. Oppenheimer seemed relieved at this news, and the two men then entered the steamy ranch house. Hubbard asked for a few minutes alone in the weather room to look over the latest data, but he found no change of significance.

A few moments later, Groves, Farrel, Don Yates, Holzman, Tolman, and Oppenheimer came in. Groves opened the conversation: "What the hell is wrong with the weather?" Hubbard replied that he had been against this date from the very first that he had heard of it. Then he turned to Oppenheimer and asked if the test could be held up as late as 9:00 A.M. Oppenheimer said that it could.

Groves then asked when the storm would clear. Hubbard began explaining that a night rain in a tropical air mass behaved differently from a standard southwestern afternoon storm. In the latter, energy from the heating of the land mass fuels the storm system, and it collapses only when the sun finally sets. But in this type of storm, he said, when the first rays of the sun reach the system, it collapses. These were the major type of storms he had dealt with during his forecasting stint in Buenos Aires. Groves said that he didn't want details, he wanted a specific time. Hubbard replied that he was giving both the details and the time so that Colonels Yates and Holzman, both meteorologists, would know if his reasoning were right or wrong. It was Hubbard's conviction that Groves was preparing to cancel the test on the spot.

Hubbard then said that Groves could postpone the test if he wished but that the weather would become acceptable at dawn.

Oppenheimer then stepped in and tried to calm the general down. He said that the project had secured the best forecaster they could find and that he believed that they should go ahead according to the forecast. Tolman, Yates, and Holzman all nodded their agreement.

Groves retreated ungraciously. You had better be right on this, he said to Hubbard, "or I will hang you." Groves insisted that Hubbard sign his forecast. By 2:45 the conference was over. The time for the shot was now set for 5:30 A.M.[27]

Immediately afterwards, Groves placed a phone call to New Mexico's governor, John Dempsey, presumably getting him out of bed. Groves had met with Dempsey several weeks earlier, and, without telling him the details, had informed him that it might be necessary forcibly and suddenly to evacuate a number of people from the Trinity region.[28] Now Groves informed the governor that he might have to declare martial law throughout the central region of the state. Mindful of the infamous Bataan death march, which had involved so many New Mexicans, Governor Dempsey agreed. Groves had also made elaborate arrangements with Mrs. Jean O'Leary, his administrative assistant in Washington, to activate any of several messages, as soon as he telephoned the proper code word. These provided for declaration of martial law over a much wider area.[29]

Hubbard then phoned Bainbridge, who was waiting at the base of the tower, and Stafford Warren to tell them of the 5:30 decision. At 3:15 Hubbard left for S-10,000 where he found that the rain had stopped. At 4:00 A.M., the wind readings began to swing around to the desired directions. T. C. Alderson of the weather group, who was stationed at the tower, reported that the clouds were breaking above him and that an occasional star had been visible from about 3:30 on. At 4:15 Hubbard completed his final forecast. At 4:40 he called Bainbridge to tell him that the winds were steady and within acceptable limits. He assured him that the atmosphere was clearing and that everything was stable enough to test. Bainbridge said that he and the others would leave the tower in twenty minutes. At 5:08, Bainbridge and Bush arrived at S-10,000 where Hubbard had his forecast in writing and signed. After Bainbridge examined it, he gave word to set the timing mechanism into operation. The test was on.

If one examines the list of desired ideal weather conditions gathered from the group leaders and compares them with the actual conditions at the time of the July 16 shot, the contrast is striking. Almost none of the optimum requirements were met. The skies were not clear; the humidity was around 80 percent, and three layers greater than 95 percent existed between 12,000 and 20,000 feet. Rain had fallen in the last twelve hours. There were three inversion layers, very slight ones at 100 feet and 500 feet, and a slight one at 17,000 feet. The only requirement that was perfectly met, however, was the wind pattern. In the end that proved most crucial.

CHAPTER FIVE

The Blast

As July 16 approached, the tension in Santa Fe and in Los Alamos almost matched that at Trinity. From her office at 109 E. Palace, Dorothy McKibbin noted that the scientists' voices over the phone became strained and taut.[1] On the Hill, husbands became noticeably edgy and disappeared for longer periods of time. The wives began to worry silently about a possible malfunction. Just before her husband was preparing to leave for the Trinity test, Elsie McMillan asked him what might happen. "We ourselves are not absolutely certain what will happen," Ed McMillan, one of the discoverers of plutonium, said. "In spite of calculations, we are going into the unknown. We know that there are three possibilities. One, that we all [may] be blown to bits if it is more powerful than we expect. Two, it may be a complete dud. Three, it may, as we hope, be a success, we pray without loss of any lives."[2] Lois Bradbury joined Elsie McMillan and both worried throughout the night. Several of the women, including Ruby Wilkening, journeyed to a nearby high ridge where they waited until dawn, listening to the shortwave radio that had been tuned to the Trinity frequency.

These wave lengths, by chance, were identical with those of a railway freight yard in San Antonio, Texas. During the final preparations, the scientists could hear the trainmen shifting cars, and, presumably, the trainmen could also hear the scientists. The ground-to-plane frequency turned out to be identical with that of the Voice of America. Anyone listening to Voice of America after 6:00 A.M.—when it came on—could also hear the Trinity scientists.[3]

On the night of July 15, a group of about ninety Los Alamos people loaded into three buses, three cars, and a truck and set off for Albuquerque. There they rendezvoused with numerous other people, including William Laurence of the *New York Times*, Sir James Chadwick, discoverer of the neutron, Ernest O. Lawrence, inventor of the cyclotron, and Edwin H. McMillan. Their goal was a hill about twenty miles to the northwest of Zero which had been dubbed "Compañia Hill." There a large group who were not immediately connected with the on-site experiments could observe the explosion. All airports within a hundred miles were told to ban aircraft from the area and they complied.

Under the direction of Norris Bradbury, the final practice assembly of the blocks of HE had been completed in Los Alamos—everything but the fissionable materials. The assembly team had practiced for weeks, and after everything was perfectly in place, the device was disassembled, packed carefully in several cars, and conveyed to Trinity. Once there, the crew moved to the base of the tower to reassemble. Each of the nearly one hundred blocks of explosives was identified, and while they fit together snugly, they were not exactly interchangeable.[4]

On July 12, the plutonium core was taken from the Los Alamos vault and carried in two separate cars to Trinity. The army officer in charge, Lieutenant Vaughan Richardson, demanded a receipt for it. Bainbridge felt this was nonsense, but General Thomas Farrell eventually signed for it. Now the valuable plutonium had been officially transferred from the army to the scientists.[5]

Preliminary assembly of the plutonium core began on the night of July 12 in the McDonald farmhouse. The scientists worked out of a box of instruments that were identical to those that had already been sent to the island of Tinian in the Pacific. On Saturday, July 14, Robert Bacher of Cornell drove to Zero with the core, where the final assembly occurred on a canvas-enclosed flooring at the base of the tower. Every step of the procedure had been carefully rehearsed so there would be no foul-ups. But, to everyone's dismay, the precision-tooled core refused to snap into place amidst the shaped explosives. The scientists hurriedly discussed the problem, but Bacher said that the nuclear material had probably just expanded slightly from the heat. He was proven right, for about five minutes later the core clicked perfectly into place. Just as they finished, a small dust storm blew over the tent, but the delicate assembly had been closed and no dirt could get in.[6]

Later that day, Norris Bradbury supervised the hauling of the bomb, still without detonators, to the top of the tower. After the hoist had lifted it into the air, someone put several GI mattresses under it for protection. This was purely psychological, however, for the mattresses could hardly have made much difference if the hoist had broken. The bomb arrived at the top of the tower, where it rested on its own base. Preparation continued throughout that day and into the next. The detonators were installed late on the night of July 15. Bradbury's log entry for that date read: "Look for rabbit's feet and four leafed clovers. Should we have the chaplain down here?"[7]

Two days before the test date, Ed Creutz sent word down from Los Alamos that last-minute tests by Ed McMillan of scale models of the gadget had been showing poor results. Judging from the figures, something was wrong with the lens system. If this were true, the test would surely fizzle. When this news arrived, numerous officials angrily grilled George Kistiakowsky on the cause of the probable failure. Oppenheimer became so emotional that Kistiakowsky impulsively bet him a month's salary against ten dollars that the implosion system would work as planned. On Sunday morning, Hans Bethe reviewed the test results carefully and concluded that the conductivity of the explosive gas that was inside the scale model had distorted the results. Bethe decided to rely on theory rather than on these last-minute experiments.[8]

Because Groves had a mania about possible sabotage, Lieutenant Howard Bush stood guard under the tower all night, much of it in the light rain. Bush must have looked at the numerous lightning safety devices with respect. A short while earlier, a model of the X unit (designed to fire the detonators), placed on the tower for a practice run, had fired spontaneously in a storm. This unforeseen event led the men to recheck all lightning protection measures—lest the atomic age be inaugurated before everyone was fully ready.[9]

When Hubbard gave Bainbridge the OK for a 5:30 firing, Bush, Joe McKibben, Kistiakowsky, and Bainbridge threw a number of relays, switches, and contacts at the base of the tower. They were the last to leave. McKibben and Bush did not go until twenty-five minutes before the time set for the detonation.[10] Contrary to rumor, they drove away at normal speed. Had there been a vehicle breakdown, they probably could have run the five miles in twenty-five minutes. Fortunately, they did not have to.

With twenty minutes to go, Sam K. Allison of the University of Chicago began counting over the loudspeaker. This became the world's first countdown. Thirty-two-year-old Joe McKibben, as a member of the "arming party" at S-10,000, threw the last set of switches, with the final one at T minus forty-five seconds. From then on until the explosion, the detonation mechanism was motorized. If the instruments had shown that something was going drastically wrong in those last forty-five seconds, however, Don Hornig could have thrown yet another switch to stop it.[11] As the countdown approached the end, a local radio station began broadcasting on the same wave length. The opening bars of Tchaikovsky's *Nutcracker Suite* strangely overlay the "nine, eight, seven. . . ."[12]

At 5:25 a rocket blasted into the sky as a signal that there were five minutes left. Another rocket was released at 5:28 as a two-minute warning, but that was a dud and sputtered out. At 5:29 a different color rocket went off, indicating there was one minute to go. It was the longest minute the scientists ever spent. Almost nobody spoke. At South 10,000 James B. Conant leaned over to Groves and whispered, "I never realized seconds could be so long."[13]

Twenty miles away on Compañia Hill, Edward Teller and Hans Bethe passed around a bottle of suntan lotion they had brought to protect against possible sunburn from the blast. All the men at the three shelters and the MPs in various distant trenches had the same instructions. They were to lie with their feet facing the blast, and they were not to look at it until after the initial flash. The men at Base Camp—ten miles away—remained in the open.

Everyone had been instructed not to look directly at the initial blast but wait a few seconds and then peer through a piece of dark welder's glass. Physicist Robert Krohn had earlier become momentarily annoyed at all the precautions and asked Enrico Fermi if such nonsense were really necessary. Fermi replied that, after all, this was an important event, and it would be a shame to miss it (by being blinded). Everyone except Richard P. Feynman did as told. Feynman became temporarily blinded in one eye.[14]

Others waited at various points around the state. Dorothy McKibbin sat in a car on the top of Sandia Peak near Albuquerque. Don Leet waited on the road near Albuquerque, and Elizabeth and Al Graves were operating instruments about thirty miles away in a Carrizozo trailer camp. Elizabeth Graves, seven months pregnant, was sent there because no women were allowed at Trinity. In San Antonio, New Mexico, MPs woke up José Miera, owner of a popular bar and restaurant, and told him that if he came outside he would see something no one had ever seen before.[15]

On Monday, July 16, 1945, at 5:29:45 A.M., Mountain War Time, the bomb ignited. The explosion created a brilliant flash that was seen in three states. It lit up the sky like the sun, throwing out a multicolored cloud that surged 38,000 feet into the atmosphere within about seven minutes. As the cloud rose, observers noticed large objects skyrocketing down from its lower third. For over an hour the immediate area lay covered with a pall of smoke.[16] The herds of antelope darted off at full speed, and no one has yet guessed where they might have first paused. The heat at the center of the blast approximated that at the center of the sun, and the light created equaled almost twenty suns. At ten miles away people felt a blast of heat equivalent to standing about three feet from a fireplace. Where the fireball touched the ground, it created a crater half a mile across, fusing the sand into a greenish gray glass that was later termed *atomsite* or *trinitite*. Every living thing within the radius of a mile was annihilated—plants, snakes, ground squirrels, lizards, even the ants. The stench of the death lingered about the area for three weeks.

Perhaps the most famous anecdote of the test involved Enrico Fermi, who was at S-10,000. Just before the final seconds, Fermi began tearing up paper into small pieces. After he saw the flash, he dropped the paper from his hand, and when the blast wave hit, measured how far it carried them. Then he pulled out his slide rule—an instrument that one scientist suggested should probably be placed in the Smithsonian Institution—and calculated the yield. He said it was in the range of 10,000 tons of TNT, which was not far from the mark.[17]

Several hundred people saw the explosion. A sheep herder for Holm Bursum lay sleeping on a cot about fifteen miles from Zero when he was awakened by the flash and blown off his cot. Bursum's foreman, Julian Jaramillo, had just saddled up his horse when the blast occurred. The horse fled into the hills, and it took Jaramillo two hours to catch it.[18]

Ranger Ray Smith, on duty near Lookout Mountain tower, northwest of Silver City, felt certain he had experienced an earthquake, as did several people in Carrizozo. The Smithsonian Observatory on Burro Mountain confirmed a shock but noted that the vibrations were unlike any earthquake ever recorded. Some observers felt they had witnessed a plane crash. The consensus in Roswell was that the explosion came from "the descent of a meteor." Eight-year-old Thomas Treat of Deming was sleeping on his front porch when the light awakened him and the roosters began crowing. He ran for his Methodist parents, and they solemnly considered if this might be the end of the world. Lewis Farris of Carrizozo ran up the main street of town shouting, "Hell's broken out someplace. Maybe it's the Japs."[19]

Monitors John Magee and Joseph Hirschfelder drove through the region following the radioactive cloud. About twenty-five miles from Zero, they rode past a mule that must have looked directly at the explosion. Its jaws were wide open, its tongue was hanging out, and it seemed temporarily paralyzed. Later it ran away. At a crossroads store, an old man remarked, "You boys must have been up to something this morning. The sun came up in the west and went on down again."[20]

Ed Lane, an engineer on the Santa Fe Railway, was in Belen when the blast occurred. He remarked later that he had a front seat for the greatest fireworks show he had ever witnessed:

> I was coming to El Paso. My engine was standing still. All at once it seemed as if the sun had suddenly appeared in the sky out of darkness. There was a tremendous white flash. This was followed by a great red glare and high in the sky there were three tremendous smoke rings. The highest was many hundreds of feet high. They swirled and twisted as if being agitated by a great force.

The glare lasted about three minutes and then everything was dark again, with dawn breaking in the east.

Mrs. H. E. Wieselman had just crossed the Arizona–New Mexico line en route from California when she saw it. She remembered,

> We had just left Safford, and it was still dark. Suddenly, the tops of high mountains by which we were passing were lighted up by a reddish, orange light.
> The surrounding countryside was illuminated like daylight for about three seconds.
> Then it was dark again.
> The experience scared me. It was just like the sun had come up and suddenly gone down again.[21]

William Hartshorn was piloting one of the two B-29s that had been sent aloft from Kirtland Air Base to track the cloud. "We didn't know exactly what to expect," he recalled, "but we didn't have to be told that huge mushroom cloud boiling up was what we had been waiting for."[22]

Yet perhaps the most powerful statement about the blast came from Georgia Green of Socorro. A University of New Mexico music student, she was being driven up to Albuquerque for her nine o'clock class by her brother-in-law. "What was that?" she asked. This might not be unusual except that Georgia Green was blind.[23]

Army security had the unenviable task of keeping this out of the newspapers. "My God," complained one security official, "you might as well try to hide the Mississippi River." They had carefully prepared themselves, however, and had several contingency releases ready, depending on the extent of destruction and the possibility of loss of life. They immediately presented the appropriate release—for New Mexico only—to all the area newspapers. Ostensibly coming from the commanding officer of the Alamogordo Air Base, William O. Eareckson, it read:

> Several inquiries have been received concerning a heavy explosion which occurred on the Alamogordo Air Base reservation this morning.
> A remotely located ammunition magazine containing a considerable amount of high explosives and pyrotechnics exploded.

There was no loss of life or injury to anyone, and the property damage outside of the explosives magazine itself was negligible.

Weather conditions affecting the content of gas shells exploded by the blast may make it desirable for the Army to evacuate temporarily a few civilians from their homes.[24]

New York Times reporter William Laurence had also prepared several contingency press releases. These were safely filed in New York. One dealt with no loss of life or property; the second discussed severe damage to property; the third detailed the obituaries of all the famous men in the immediate area, including himself. Laurence had the Tom Sawyer–like experience of writing his own obituary. He enjoyed concocting this version of the explosion—all the people supposedly died from a freak accident at Oppenheimer's ranch in the Pecos Mountains. As he recalled years later, he "out Roger'd Buck Rogers" and "out Wells'd H. G. Wells."[25]

As ever, General Groves was terribly concerned about secrecy. He even wanted the scientists to keep the explosion from their wives. Security all across the nation had been alerted that something big was in the offing. If things went wrong, they knew they would have to take emergency action.[26] Even without emergency action, they still had their hands full.

The War Department asked the local newspaper editors that they use no background material, add no details, and make no explanations. Most of them complied and printed all or part of the official statement. Enough people had seen the blast to make this a page-one headline, but only the El Paso *Herald Post* did so. Those papers that held back, such as the Silver City *Press*, were annoyed at the extensive coverage allowed the *Herald Post*. Several people reported the flash to the Albuquerque *Journal*, but security pressured them not to cover the story in detail.[27]

The arm of security proved a long one. A cub reporter for a Chicago paper received a phone call from a man who had been traveling through the area by train. He told her, in great detail, about the crash of a huge meteor. So, she wrote a short article on it. The next day she was called into her editor's office and grilled by two FBI agents until she promised not to write any more on it.[28] The story also broke on West Coast radio, but it was quickly silenced.

The official explanation of the ammunition dump satisfied people from the relatively distant areas such as Albuquerque.[29] Many local residents, however, remained dubious. "Everybody knew something had happened," recalled Beatrice McKinley of Alamogordo. "The stories they told were very clumsy. . . ." The Alamogordo *News* printed the official release but also reported the local consensus that "some experimentation was going on in explosives which required an isolated terrain such as the explosion occurred on." A nearby Lincoln County woman, however, dismissed the whole story as nonsense. She said that she knew a bomb when she observed one. She could hardly miss it, she said, when it shook the whole house and threw things down from the top of her kitchen cabinet. Her comments, however, were not printed until after the war was over, when the nuclear cat was safely out of the bag.[30] People in the area were still discussing the effects of the blast several months later.[31]

If the lay reaction bordered on awe, the reaction of those in the know was even greater. Meteorologist Jack Hubbard quickly felt himself to make sure he was still alive. Physicist Marvin Wilkening was struck by the dazzling colors, the purples and oranges at the head of the cloud as it surged into the atmosphere. Photographic expert Berlyn Brixner likened it to the turning on of a gigantic searchlight right in front of his eyes. I. I. Rabi recalled that the light seemed to bore its way into you—so much so that you wished it would stop. The light, remembered Norris Bradbury, "was beyond belief in terms of any other thing I'd seen."[32]

"It beggars description," remarked seismologist L. Don Leet of Harvard:

> I was fifty miles from the explosion awaiting a message. There were thunderstorms on the horizon and I was afraid that I would not see the flash.
> When it let go, it lit up 180 degrees of the horizon, not like one but a dozen brilliant suns. It stayed lit up and made chills run up my back because I knew what might happen if it was not controlled.
> It was followed by a brilliant red wall of flame. Fifty miles away it was like an earthquake. At the source the brilliance was so great that an individual looking at it would be blinded. One observer did so and was blinded.
> It created a crater half a mile across and a quarter of a mile long.

The seismic disturbance caused does not equal that of an earthquake but the force is so concentrated that I would rather go
through an earthquake than face that explosion.[33]

While the dazzling lights at the head of the cloud—caused by
the ionization of the air—were the most startling, several observers were equally impressed by the sound. SED Leo M. (Jerry)
Jercinovic recalled the growing rumbles as the sound waves
bounded across the valley at fifteen-second intervals before they
gradually merged into a low, steady rumble. It was strong enough
that the entire valley seemed to shake. Roy Carlson was over
twenty miles away, but he felt as if he were right next to it. Jack
Hubbard at S-10,000 likened it to an express train that had passed
within inches.[34] "I can still hear it," wrote Otto Frisch in 1979.[35]

General Groves spent the next few days compiling his impressions in a memorandum to be sent to Secretary of War Henry L.
Stimson at Potsdam. While Groves's observations are not lacking
in perception, he fortunately also included those of his chief
assistant, Brigadier General Thomas F. Farrell. Farrell possessed
a genuine literary flair:

> In that brief instant in the remote New Mexico desert the tre-
> mendous effort of the brains and brawn of all these people came
> suddenly and startlingly to the fullest fruition. Dr. Oppenheimer,
> on whom had rested a very heavy burden, grew tenser as the last
> second ticked off. He scarcely breathed. He held on to a post to
> steady himself. For the last few seconds, he stared directly ahead
> and then when the announcer shouted "Now!" and there came
> this tremendous burst of light followed shortly thereafter by the
> deep growling roar of the explosion, his face relaxed into an expres-
> sion of tremendous relief. Several of the observers standing back
> of the shelter to watch the lighting effects were knocked flat by
> the blast.
>   . . . All seemed to feel that they had been present at the birth
> of a new age—the Age of Atomic Energy—and felt their profound
> responsibility to help in guiding into right channels the tremen-
> dous forces which had been unlocked for the first time in history.[36]

Perhaps the most lyrical description, however, came from the
pen of William L. Laurence:

> The Atomic Age began at exactly 5:30 Mountain War Time on

the morning of July 16, 1945, on a stretch of semi-desert land about 50 airline miles from Alamogordo, N.M., just a few minutes before the dawn of a new day on that part of the earth.

Just at that instant there rose from the bowels of the earth a light not of this world, the light of many suns in one. It was a sunrise such as the world had never seen, a great green super-sun climbing in a fraction of a second to a height of more than 8,000 feet, rising ever higher until it touched the clouds, lighting up earth and sky all around with a dazzling luminosity.

Up it went, a great ball of fire about a mile in diameter, changing colors as it kept shooting upward, from deep purple to orange, expanding, growing bigger, rising as it was expanding, an elemental force freed from its bonds after being chained for billions of years.

For a fleeting instant the color was unearthly green, such as one sees only in the corona of the sun during a total eclipse. It was as though the earth had opened and the skies had split.

One felt as though he had been privileged to witness the Birth of the World—to be present at the moment of Creation when the Lord said: "Let There Be Light."[37]

A variety of emotions cascaded through the mind of the director, J. Robert Oppenheimer. According to Hubbard's diary, around 6:30 A.M. Oppenheimer remarked that "my faith in the human mind has been somewhat restored."[38] Afterwards he told a group of reporters that while he was "greatly relieved" that the bomb worked, he also admitted that he was a "little scared of what we had made."[39] He also confessed that through his mind had flashed the line from the Bhagavad Gita: "I am become Death, the destroyer of worlds."[40]

Most of the observers were highly articulate, but even they struggled to put what they had heard, seen, and felt into words. James Tuck of the British Mission asked with a gasp, "What have we done?" Others whispered, more in reverence than otherwise: "Jesus Christ." An assistant to Julian Mack at N-10,000 said, "My God, it's beautiful." "No," replied Mack, "it's terrible." "I knew the world would never be the same again," said Dorothy McKibbin, "and it hasn't been." George Kistiakowsky felt this was the nearest thing to doomsday that one could imagine. "I am sure," he said, "that at the end of the world—in the last millisecond of the earth's existence—the last men will see what we saw."[41]

Yet of all those who recorded their impressions, none has matched the wisdom of I. I. Rabi:

> At first I was thrilled. It was a vision. Then a few minutes afterward, I had goose flesh all over me when I realized what this meant for the future of humanity. Up until then, humanity was, after all, a limited factor in the evolution and process of nature. The vast oceans, lakes and rivers, the atmosphere, were not very much affected by the existence of mankind. The new powers represented a threat not only to mankind but to all forms of life: the seas and the air. One could foresee that nothing was immune from the tremendous powers of these new forces.[42]

Generally speaking, however, the philosophical observations came later. The initial reaction was: it worked! The two billion dollars and the years of effort had not been wasted. Both William Laurence and General Farrell were amazed at how the explosion momentarily turned the distinguished scientists into little children. George Kistiakowsky, who had actually been thrown to the ground by the blast and was covered with mud, rushed up to Oppenheimer and hugged him. At South 10,000 a group formed a spontaneous chorus line and danced around in a snake dance. One after another the scientists got on the PA system and began howling jubilantly.[43]

"Some people claim to have wondered at the time about the future of mankind," remarked Norris Bradbury. "I didn't. We were at war and the damned thing worked."[44] Marvin Wilkening, then a young graduate student with Fermi's group, was fascinated with how relatively simple it all had been.[45] Ernest O. Lawrence, watching from Compañia Hill, felt the same way. What impressed him most was that the whole affair had gone off exactly as the scientists' calculations had predicted.[46] It was, in the parlance of the time, "technically sweet."

Test director Kenneth Bainbridge also sought out Oppenheimer to congratulate him. His comments combined both philosophical reflection and the reaction of the moment. "Well," Bainbridge said to Oppenheimer, "now we're all sons of bitches."[47]

If the initial response was elation—the group on Compañia Hill had actually applauded—the aftermath brought on a different mood. When the scientists piled into the buses and cars to return to Los Alamos, they became genuinely solemn. Ernest O.

Lawrence and William L. Laurence both felt the experience bordered on religious reverence.[48] But many of the men were uneasy by what they had accomplished. No one phrased it better than Victor Weisskopf: "Our first feeling was one of elation, then we realized we were tired, and then we were worried."[49]

The scientists immediately involved, of course, were exhausted, and most of them made their way back to the Hill as best they could. But security restrictions still remained uppermost in many people's minds. When one carload of exhausted men stopped at a small restaurant in Belen for a quick bite, they pretended not to notice a second, identical carload that arrived a few minutes later.

Enrico Fermi was so drained by the experience that he asked another person to drive his car, something he almost never did. When he returned to Los Alamos, his wife, Laura, described him thus: "he seemed shrunken and aged, made of old parchment, so entirely dried out and browned was he by the desert sun and exhausted by the ordeal."[50]

Several observers at Los Alamos had seen the flash, but as they lacked details, they spent the morning worrying. Thus, a crowd of about forty people gathered at the Los Alamos radio station transmitter to hear the 12:30 news from KOB in Albuquerque. When they heard the "official" statement about the ammunition dump explosion, they breathed a sigh of relief. When the scientists began to stagger in to the East Cafeteria in Los Alamos around 6:00 P.M., July 16, they were embraced by their wives and friends and peppered with questions.[51]

Stan M. Ulam, who had decided not to accompany the caravan to Compañia Hill, joined the welcoming group. "Somehow I didn't feel like going," he confessed. "It was purely nervous or psychological. A sort of block, if you want to call it that." Yet he remembered watching the buses return: "You could tell at once they had had a strange experience. You could see it on their faces. I saw that something very grave and strong had happened to their whole outlook on the future."[52] Only with Trinity did the scientists comprehend the full potential of the forces they had unleashed. The power of the fissioned atom was greater than anyone had ever imagined.

Hauling Jumbo to Trinity Site.

Unloading Jumbo at Pope's Siding.

Jumbo resting on its base at Trinity Site.

Jumbo, undamaged, immediately after the Trinity explosion.

Jumbo lying in the desert. (Courtesy of the Socorro Historical
Society.)

Louis H. Hemplem

Above, left: Louis Hempelmann, with his name misspelled, from a badge photo.

Left: Stafford Warren. (Courtesy of Mrs. Stafford Warren.)

Above, right: James Nolan. Nolan, Hempelmann, and Warren were the three physicians who established the health safety procedures for the Trinity shot. (Courtesy of Mrs. James Nolan.)

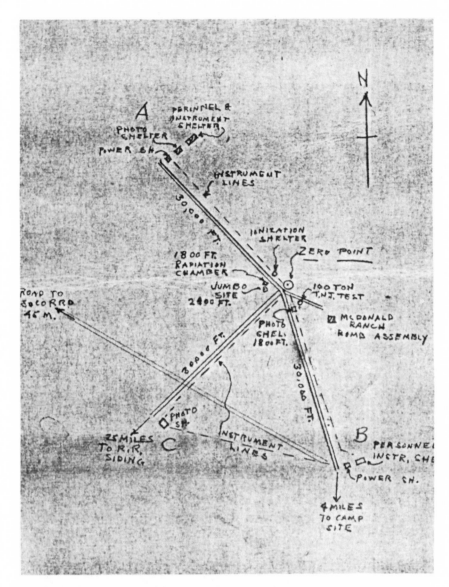

Parchment map carried by Sam Davalos when he supervised the construction at Trinity Site. (Courtesy of Sam Davalos.)

Stacking the TNT for the 100-ton shot. (Courtesy of LANL and the National Atomic Museum.)

Base Camp at Trinity Site. (Courtesy of LANL and the National Atomic Museum.)

Main instrumentation and firing bunkers at Trinity Site. (Courtesy of LANL and the National Atomic Museum.)

Equipment inside the shelter at N-10,000.

The tower for "the gadget."

"The gadget" resting on top of the 100-foot steel tower at Trinity Site. (Courtesy of LANL and the National Atomic Museum.)

One of the tanks used to enter the blast area immediately after the explosion.

My Observations During the Explosion at Trinity on July 16, 1945 — E. Fermi

On the morning of the 16th of July, I was stationed at the Base Camp at Trinity in a position about ten miles from the site of the explosion.

The explosion took place at about 5:30 A.M. I had my face protected by a large board in which a piece of dark welding glass had been inserted. My first impression of the explosion was the very intense flash of light, and a sensation of heat on the parts of my body that were exposed. Although I did not look directly towards the object, I had the impression that suddenly the countryside became brighter than in full daylight. I subsequently looked in the direction of the explosion through the dark glass and could see something that looked like a conglomeration of flames that promptly started rising. After a few seconds the rising flames lost their brightness and appeared as a huge pillar of smoke with an expanded head like a gigantic mushroom that rose rapidly beyond the clouds probably to a height of the order of 30,000 feet. After reaching its full height, the smoke stayed stationary for a while before the wind started dispersing it.

About 40 seconds after the explosion the air blast reached me. I tried to estimate its strength by dropping from about six feet small pieces of paper before, during and after the passage of the blast wave. Since at the time, there was no wind I could observe very distinctly and actually measure the displacement of the pieces of paper that were in the process of falling while the blast was passing. The shift was about 2½ meters, which, at the time, I estimated to correspond to the blast that would be produced by ten thousand tons of T.N.T.

Enrico Fermi's observations on the Trinity blast.

Ground Zero two
days after the
explosion.
(Courtesy of
LANL and the
National Atomic
Museum.)

28 HOURS
VERTICAL   NORTH   100

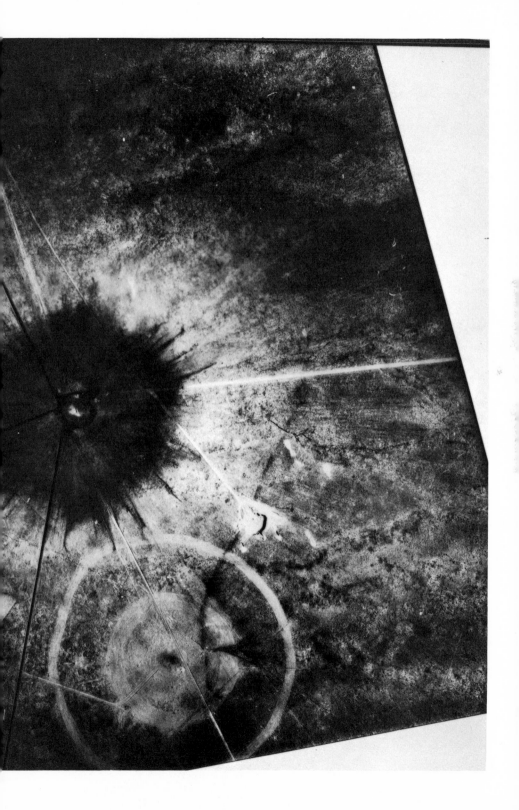

Recovery teams and radiation monitors.

Visitors to Ground Zero eight weeks after the explosion. (U.S. Army photo, courtesy of the Albuquerque *Journal*.)

Fallout-damaged cattle and their offspring at Oak Ridge. Photo courtesy of Atomic Energy Commission, Oak Ridge, Tennessee.

The monument at Ground Zero.

The plutonium bomb, "Fat Man," on the island of Tinian. (Courtesy of LANL and the National Atomic Museum.)

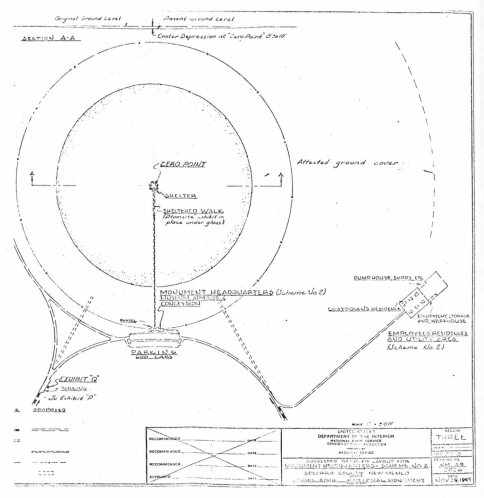

The 1945 Park Service plan for an atomic bomb national monument at Trinity Site. Note the naive attitude toward radioactivity at Ground Zero.

114

CHAPTER SIX

The Aftermath I: Fallout

Two Trinity observers not jumping up and down at the moment of detonation were Stafford Warren and Louis Hempelmann. As physicians in charge of health safety for the site, their job was still to come. "If [the cloud] went east," Warren later wrote, "I was sunk. If it went northeast, it was going to be all right." "All I could think of," Hempelmann recalled, "was, my God, all that radioactivity up there has got to come down somewhere."[1]

The radioactive cloud split into three parts within about fifteen minutes. The lower section drifted north, and the center one west. The brunt of the cloud moved toward the northeast at 45,000–55,000 feet at approximately 10–14 MPH. It drifted in the direction of the towns of Adobe, Claunch, Encino, and Santa Rosa. Very little fallout reached the ground for about two hours, however, and this gave everyone a false sense of security. Then the cloud dropped a swath of fairly high radioactivity over an area roughly a hundred miles long by thirty miles wide. Ground-level readings of 3.3 Roentgens (R) per hour became common. Later that day, traces of gamma radiation showed up in Santa Fe, Las Vegas, Raton, and even Trinidad, Colorado, which is 260 miles away. Enrico Fermi estimated that the radioactivity emitted by the blast was equal to one million times that of the world's total supply of radium.

While the two B-29s from Kirtland had been prevented by the weather from dropping their pressure gauges near the explosion, they were able to follow the gigantic cloud for several miles.

They lost it only when it began to merge with the surrounding thunderheads. Other planes, equipped with special filters to measure radioactive intensity, followed the cloud across the nation. Jack Hubbard piloted one trace plane for several hours and then estimated where the cloud would go from there. Based on the Krakatoa explosion of 1883, he guessed that it would take thirty-six hours to go around the world. Because it was summer, the cloud moved slowly. From New Mexico it drifted across Kansas, Iowa, Indiana, upstate New York, New England, and, finally, out to sea. When it came around the earth again, no traces of activity could be found.[2] This was America's first acquaintance with radioactive fallout. It introduced questions which, almost forty years later, are still not completely answered.

Fortunately, the wind structure was optimum for the test. The early cloud moved over uninhabited regions with no population center in its path. Few people in the vicinity were exposed to heavy radiation.

The land under the tower, of course, was most affected by the radioactivity from the blast. The scientists planned to enter the region as soon as possible to gather samples of the soil. They were convinced that dirt samples taken from the center of the crater would allow them to calculate the blast efficiency, the percentage of plutonium that had undergone fission.

At one time they considered sending a small blimp or a helicopter over the crater with a winch, but they abandoned that as too risky. Moreover, 1945 helicopters did not fly well above 5,000 feet. Instead, they decided that the best entry vehicle would be a tank. The army provided two T-4 tanks. A white one was lined with two inches of lead—twelve tons total—while a silver one was also equipped with a rocket launcher for taking samples at a distance. Each vehicle carried its own air, as well as special devices for sampling the soil.

Immediately after the blast, the tank with the winch started for the crater, but suddenly the engine died. Several physicists rushed over to try to repair it on the spot, but to no avail. Fortunately, it had only traveled 500 yards and had not reached the contaminated area when this occurred. Shortly afterwards, the problem was corrected.

Enrico Fermi had planned to join the tank crew for the first trip into the crater, but he became ill and declined the honor. This left Herbert L. Anderson in command. Anderson and his driver entered the blast area around 7:00 A.M. and were the first to discover the trinitite. Anderson radioed back that the region had turned "all green." Even with the lead-lined tank, however, the radioactivity was so intense that they went in only a short way before quickly turning around. Later that day, the tanks made five forays through the crater region to gather earth samples. Two of these trips went right through the center of the crater.[3] The soil was sent to Los Alamos for analysis. After a week of testing, Anderson told an anxious Oppenheimer that the blast had an 18 percent efficiency and was equivalent to 20,000 tons of TNT. These measurements provided the figures that were announced to the public after the Nagasaki blast.[4]

Because the region around Zero was highly radioactive, the health personnel maintained a close watch on those who went in. For the first three days after the shot, when numerous people asked permission to enter, a special "Going-in Board" ruled on all requests. All those who did enter the region wore protective clothing and dust respirators. They were monitored extensively. The board tried to keep all single-day exposures below 5 Roentgens a day.

As the site cooled off, the physicians began to relax the precautions, but a careful "Trinity Exposure Record" was kept for each person.[5] Health personnel remained at Trinity from July 16 to August 14, but problems of scientific "sightseeing" were such that Louis Hempelmann had to return in mid-October to tighten the regulations regarding entry. By January 1946, as Paul Aebersold noted in his official report, the activity had decayed to such a level that a person would have to "camp" at Zero to accumulate a harmful dose of radioactivity.[6] A high chain link fence separated the crater region from the surrounding desert.

On one of these later observation trips, Victor Weisskopf drove a jeep with Bethe, Fermi, Oppenheimer, and Groves over near Ground Zero. As Groves viewed the crater, he remarked: "Is that all?" Weisskopf suggested that Groves probably expected a hole all the way to the center of the earth.[7]

Before the test, Bainbridge had worried that sufficient plutonium would be left in the soil "in the vicinity of the shot" to make it uninhabitable for some time afterward. This certainly proved true. The blast deposited an estimated 1–2 percent of the total fission products on the ground, and the induction of radioactivity into certain elements in the soil from capture of neutrons caused this area to register at 600–700 Roentgens per hour.[8] Victor Weisskopf noted that the gamma radiation also proved much more intense than they had predicted. Anyone observing the blast from the nearest anticipated "safe" position would have been killed instantly.[9] The radiation dropped off rapidly with the distance from Ground Zero, however, so that outside a radius of 1,200 yards it was much less. The beta and gamma activities decayed at the rate of $1/t$, where t was the time since fission.

Radioactivity, of course, is a fact of life for all living creatures. Some areas of the globe have a much higher natural radioactivity than others, and altitude also helps determine the levels of cosmic radiation. Granite is especially high in natural radioactivity, and Grand Central Station in New York City (built with granite blocks) became notorious as a "hot box." Even today, Park Service people who work at Carlsbad Caverns are limited in the amount of time they can spend underground. Sometimes they even wear radiation badges.

While radioactivity may be a fact of life, concepts of acceptable dosage levels have fluctuated considerably over the years. In 1934, the International Commission on Radiological Protection set the maximum permissible dose of radiation at .2 R per day. Two years later the United States National Committee on Radiation Protection and Measurements set the level at .1 R per day (36.5 per year). It was assumed that a person could tolerate .1 R a day for life. This figure remained standard during World War II. In the late 1950s, after leukemia had struck down many of the early experimenters, the dosage was reduced. Today it is around 5 R per year, although it is expressed in a different unit— a REM (Roentgen equivalent in man).[10]

It was generally understood that any excessive amount of radiation damaged the body. If the dosage were below an absolutely lethal level, however, the results were uncertain. Some people would be affected and others spared. No one could predict with certainty, for the effects could be measured only in a large statistical group. The advent of subatomic physics had changed nuclear measurements to statistical probabilities. The advent of radioactive fallout introduced the same uncertainty to those exposed to low doses of radiation.

In the period between the wars, several universities had experimented with radiation therapy, primarily as a way to treat cancer. This experimentation, however, as one observer noted, "was an empirical speciality rather than one based on precise fundamental knowledge."[11] Still, by 1942 the researchers had created a considerable body of literature on radiation. These studies formed the basis for many of the safety procedures of the Manhattan Project.

The general public received its first knowledge of the dangers of radiation from the publicity during the late 1920s on the radium dial workers. Beginning in World War I, several groups of workers, usually young women, were assigned the task of painting clock and instrument dials with radium paint. These people often licked their brushes and consequently ingested significant amounts of radium. The radium proved soluble and was soon deposited in the bones of the workers. Many of them suffered gross disfiguration from fractures and bone cancer, or even death. The case was widely known.[12]

During the post–World War I era, radium was frequently touted as a medical panacea. Facial radium applications, numerous radium injection preparations, radium salves, devices for inhaling radium emanations, and even a drink, "Radithor," sold well into the early 1930s. Eventually, they were all removed from the market. The publication of Percy Brown's *American Martyrs to Science through Roentgen Rays* (1936) signaled a growing sophistication in radiation protection.

From the onset, concern over radiation safety formed a major dimension of the Manhattan Project. Radiobiological work, later called the Plutonium Project, was carried on at several sites around the nation: the University of California, the University of Chicago, Oak Ridge, and the National Cancer Institute in Bethesda, Maryland.[13] Because of concern over security, the term *health physics* gradually emerged as the cover name for all radiation protection measures.

Related experiments were also done at the University of Rochester. In March of 1943, Rochester began a special radiological safety program under the aegis of Stafford L. Warren, who was to advise the Manhattan Project on "special materials." Part of the university's assignment was to help determine safe dosage levels for project workers. This involved ashes, wastes, hair and body contamination, dusts, and inhalation hazards. When Warren was commissioned a colonel and called to Oak Ridge, A. H. Dowdy took over. Beginning with seven people, by 1945 the Rochester project had grown to 350. One division irradiated animals to try to determine acceptable dosage levels. Two hundred monkeys, 675 dogs, 20,000 rats, 277,400 mice, 1,200 rabbits, 100 hamsters, and 50 million fruit flies were involved. There were so many experiments, Rochester had to hire its own veterinarian. Another division developed the survey instruments that were used to measure the radioactivity on the ground after the Hiroshima shot.[14] It was under the Manhattan Project that modern-day radiation protection measures developed into maturity.[15]

Health safety at Los Alamos was assigned to Louis Hempelmann. He established safe exposure levels and introduced new, complicated methods of monitoring. Throughout the first year, when the project was small, there were few health safety problems. Consequently, Hempelmann helped the post surgeon on his daily rounds. From 1944 on, however, his radiation responsibility grew, and he had to devote full time to it.[16]

While there were no fatal radiation accidents at Los Alamos during the war years, there were several close calls. Out of ignorance or bravado, many scientists were careless about their experiments.[17] Some plutonium blew up in the face of one worker, and a member of the British delegation received a high accidental dosage when he worked several hours partly inside a cyclotron. Otto Frisch was once working with some unshielded U-235 when he leaned over the assembly for about two seconds. His body reflected some of the neutrons back into the uranium and caused it to go critical. "Actually, the dose of radiation I had received was quite harmless," he noted later, "but if I had hesitated for another two seconds before removing the material (or if I hadn't noticed that the signal lamps were no longer flickering!) the dose would have been fatal."[18]

Radiation safety efforts increased dramatically in Los Alamos in February 1944, when the first significant amounts of plutonium began to arrive. Created by bombarding U-238 with neutrons, plutonium, as the first man-made element, lay full of uncertainties. Groves did not exaggerate by much when he declared that the development of plutonium was as great an adventure into the unknown as was Columbus's voyage.[19]

Hempelmann established strict rules for plutonium at Los Alamos and lectured all sections of the laboratory on its toxicity. Everyone considered the hazards of handling plutonium as comparable to those of the radium industry.[20] Given the risks, the Manhattan Project in general and Los Alamos in particular had a remarkable health safety record. Only after World War II did Los Alamos have its two fatal criticality accidents.[21]

Health safety for the surrounding (off-site) areas was also a major concern. The "suicide squad" for the first chain reaction experiment at Chicago was designed to save the city from exposure to radioactivity. Scientists at Hanford, Washington established careful monitoring systems for all water and smoke released from the plants. But the health problems arising from the Trinity explosion raised the question to another level. They now concerned the entire nation, perhaps even the world.

At the instant of detonation, 360 radioactive isotopes were created by the fission of plutonium, but their natural decay destroyed most of them before the cloud reached its maximum, and the wind took over. Strontium-90, later to become the *cause célèbre* of fallout during the 1950s, does not appear until fourteen minutes after detonation. The Trinity bomb turned out to be not terribly efficient, and it spread plutonium, deposited on dust particles, over thousands of acres. The long-term characteristics of plutonium, moreover, remained uncertain. All that was known was that its half-life was 24,000 years.[22]

Where the fireball touched the earth, it pulverized the soil, vaporizing all the upper portions of the ground—chiefly sand. It was estimated that perhaps 100–250 tons of sand were vaporized and sucked up into the central and upper portions of the ball of fire. As the blast began to cool, the silica precipitated out in the form of a fine smoke. Since the radioactive material in the fireball adhered to whatever solid it happened to hit, many of these particles, derived from volatized silica were active.[23]

Uncertain as to yield, direction, and height of cloud, and worried by the predictions of Hirschfelder and Magee, the Trinity radiologists nevertheless established the best program they could to monitor the area. A planning committee, headed by Bainbridge, drew on the expertise of Hempelmann, Warren, and Nolan. Nolan ended up organizing and directing most of their activities. At Warren's urging, they mailed out 142 registered letters to various towns all over New Mexico. The letter contained badges of Eastman Kodak film. Since they were never picked up, the postmasters eventually returned them to Los Alamos. Similarly, 90 other nonregistered letters were also sent around the state.

Most of the letters found their way back to Los Alamos, and the exposed films allowed the radiologists to calculate the amount of radiation each community had received. Nearby Bingham, for example, at 3.35 R, and Cedarvale at 8.2 registered the greatest exposures. The majority of the films, however, came back blank.[24]

The scientists also established elaborate precautions for all the men in the immediate region. Each shelter had numerous respirators, as well as a designated shelter chief and a medical officer who were to make the decisions on evacuation. Everyone involved was instructed to familiarize himself with the best escape route, and some of the major roads were regraded. Marvin Wilkening recalled that the night before the test, Fermi took all of his group aside and carefully rehearsed the evacuation plans. Each area had a specially designated monitor, who was to coordinate the escape. He was also to be the last to leave the area. Once outside the "danger" region, however, there were no specific plans. Each person would make his way back to safety the best he could. Other officials were stationed throughout the area in case the men at the shelters were all killed.

The physicians also concerned themselves with the mental health of the scientists, should the detonation not go as scheduled. The tension was incredible, and the chance of severe trauma and psychological disorientation loomed as real fears. Consequently, Stafford Warren alerted the four psychiatrists at Oak Ridge to be ready to fly to Los Alamos at a moment's notice.[25]

The scientists also established contingency plans to evacuate the inhabitants from the surrounding region. The week before the test, numerous medical doctors from Oak Ridge arrived in Los Alamos to free the local physicians to go to Trinity in case of emergency. On July 14, Major T. O. Palmer, Jr., and Major Russell B. Miller, transportation officers, took a contingent of 103 troops from Company B, stationed in Santa Fe, to the vicinity of the site. So secret was this mission that no mention of it was made in the morning report. Special orders were issued covering the rations that were not drawn that day. Armed with trucks and jeeps—140 in all—plus food and water, they were to be prepared to evacuate the surrounding area if necessary.[26] About fifty MPs were also alerted.

There was some discussion in Los Alamos about *prior* evacuation of the nearby region—Warren, especially, urged this—but that was overruled because of security problems. There would have been too many explanations to the ranchers and to the press. Moreover, the area ranchers were an independent lot, and several might have refused to leave. As a result, the scientists decided to wait and see. If certain regions had to be evacuated, Palmer, Miller, the MPs, and the troops stood ready to do so. General Farrell was also alerted to be ready should any military emergency arise.[27] Groves, for some reason, was especially concerned about Amarillo, Texas, over three hundred miles away.

The evacuation plans were flexible. If only a few ranch families were involved, the army planned simply to send them to a hotel in a nearby town. But the military also stocked Base Camp with extra tents and supplies so that they could shelter and feed 450 people for a two-day period, if necessary. If the fallout began to endanger the larger population centers, the army planned to transport the people to barracks at either Alamogordo Air Base or Kirtland in Albuquerque. The base commanders were alerted to prepare to house several hundred people, and, accordingly, they laid aside appropriate food and cots.[28]

There were several varieties of monitors readied at Trinity, all under Hempelmann's jurisdiction. Paul Aebersold was in charge of on-site monitoring, with Stafford Warren as his consultant. Joe Hoffman was listed as "roving monitor" and had charge of all off-site activities. In Albuquerque, Lieutenant Colonel Hymer Friedell maintained constant radio contact with both the roving monitors and the military. He was deliberately stationed over a hundred miles away in case an unexpected disaster struck Trinity.

On July 13, Richard Watts installed stationary recording meters at Carrizozo, San Antonio, and Hot Springs. Later, other meters were put at Socorro and Magdalena. Additional instruments were not placed farther out because of lack of information about the wind. Some of these stations were manned. In all, about forty-four people were involved in collecting fallout data after the blast.

Even more important than the stationary monitors were the moving monitors. Armed with radiation meters, they were based in Carrizozo, Roswell, Fort Sumner, Artesia, and elsewhere. Their program was flexible. Shortly after the blast, they were to call Friedell in Albuquerque to ascertain wind direction, velocity, drift, and height of the cloud. Then, depending on whether the blast were a north or south blow, they had specific routes to cover. Nicknamed "Hempelmann's Rough Riders," they used their own code, based on their ages: Joe Hoffman was "Joe 32"; Joe Hirschfelder was "Joe 34"; John Magee was "John Mac 30," and so on. Periodically they stopped to take samples of earth and readings from the air.[29] The monitors were told that they were "observers" only. All data had to be reported to the chief roving monitor, Joe Hoffman, who was in radio contact with both Friedell and Hempelmann; they were the ones "to advise on evacuation."[30] Other monitors, under Wright H. Langham, also reported to Hempelmann, who was stationed with the military test command post at S-10,000.

The only evacuation, however, came from within the operation itself. When the cloud split into thirds, the lowest section, with the heaviest particles, began to drift in the direction of N-10,000. About twenty minutes after the blast, Captain Henry L. Barnett cried out, "alphas are rising; we've got to get out of here." The men looked up and saw a large red cloud of "stuff" descending on them. Barnett's meters showed a reading of ten Roentgens per hour, and when he showed this to shelter chief R. R. Wilson, he "did not check center setting but advised immediate evacuation of everyone." Berlyn Brixner, meanwhile, was busy unloading his numerous cameras. This proved a time-consuming process because the cameras were set at different speeds, and some had not yet used up all their film. Everyone had piled into the escape vehicles, but Brixner was still unloading film. Finally several people shouted at him, "Brixner, come. We've got to get out of here." Filling his arms with cameras, Brixner hopped into the nearest truck and pulled the film out along the way. About thirty men left in three vehicles along a sand-covered road. Eventually they made their way back to Base Camp.[31]

It was never clear how much radiation these men at N-10,000

received. They were wearing radiation badges, but when the badges were developed, the results were negative.[32] On-site monitor Paul Aebersold arrived to investigate the situation about two and a half hours later. He found only very low readings (0.01 to 0.2 R per hour) and notified Base Camp that N shelter was safe for those who needed to return.[33] Later, there was talk that the shelter monitor might have accidentally caught the zero knob on his meter with his thumb while changing sensitivity scales or misread the numbers.

Around 9:00 A.M., Bainbridge called Hubbard from S-10,000 to ask if the winds had shifted. A disagreeable-looking fog was approaching them, and their Geiger counters had begun to click wildly. Several people at S-10,000 had put on dust masks. Hubbard, who had to be awakened from a sleep of exhaustion, assured Bainbridge that the upper winds were holding steady. Any rise in readings at S-10,000 and Base Camp would be due to the part of the cloud that had been trapped below a low-level inversion. Hubbard assured Bainbridge that the inversion was due to break up between 9:00 and 9:30, when the thermal updrafts would carry the clouds aloft again.[34]

The two searchlight stations, L-7 and L-8, both on Highway 380 east of Bingham, also had to be evacuated. Based on Hubbard's twelve-hour weather forecast, which proved accurate, these portable units were set there to help track and illuminate the cloud. Since the shot occurred at dawn, there was little need for illumination, but the searchlights did provide brief azimuth and elevation data before the cloud became obscured by the other clouds in the sky.[35] It was not long, however, before roving monitor Arthur Breslow urged the L-7 crew to depart. As Breslow drove east on Highway 380, he discovered, to his dismay, that he had left his respirator back at the L-7 position. Ahead of him lay a valley covered with a stratum of sandlike radioactive dust through which he had to drive. Closing the windows, he drove into it while breathing through a slice of bread. As the smoke pots had indicated, the radioactive cloud sank into the nearby valleys.

The nearby L-8 crew had tracked the cloud for one hour and ten minutes, until it became too diffuse to identify. Their job done, they built an open fire and began cooking the steaks they had saved for the occasion. When monitor Carl S. Hornsberger

arrived around 8:30, he noted readings of 2 R per hour and climbing. The crew reluctantly buried their steaks in the ground and
departed.[36]

The monitors reported other readings from Roswell, Bingham,
Carrizozo, and Vaughan. None came from Socorro because the
automatic recording device ran out of red ink. One monitor, who
was eleven miles west of Vaughan on Route 60, found that his
meter suddenly jumped off scale. It indicated readings of 7–14 R
per hour. On the fourth day after the explosion, other monitors
still found parts of Vaughan where the meters would do this. At
8:49 A.M., a monitor reported a reading of 8 R per hour on the
road four miles east of Bingham. Hempelmann was notified and
ordered continued monitorings on US 380 east of the town.

The Vaughan and Bingham readings both reached such high
levels that serious thought was given to evacuating the two
towns. The scientists decided against this only because the readings dropped rapidly once the large radioactive cloud had passed
over head. The readings from the ground were easily measurable,
but they were deemed within acceptable limits. Kenneth Bainbridge recalled, however, that one family off Highway 380, between Socorro and Carrizozo, received enough fallout on their
land so that the military requested them to stay indoors for a
few days until the fallout had declined to a safe intensity.[37]

John Magee was also monitoring the Bingham area. Around
8:30 A.M., he reported a reading of 20 R per hour for an area 3.5
miles to the east and south of L-8. This was ever-after known as
"Hot Canyon."[38] Twenty R was a very high reading, and he immediately radioed it in to Hempelmann. Hempelmann recorded
the figures and instructed Magee to take another meter and
remeasure the area. Then he lost radio contact with the monitors. No one had thought to test the frequency range for the
various car radios. "It was a very frightening moment," Hempelmann recalled.[39]

When this reading became known, Groves grounded all observation aircraft heading toward the cloud, and Friedell was
called in from Albuquerque. Hempelmann and several others
immediately drove to the canyon, which lay about twenty miles
from Ground Zero. Magee meanwhile checked with Joe Hoffman
about his readings. He thought to himself that the canyon had
probably become a disaster area. In 1982, Magee confessed that

he still considered his discovery of Hot Canyon the most dramatic thing that had ever happened to him.[40] By the time that reinforcements began to arrive in the area, however, the cloud had passed overhead, and the readings started downward. The canyon proved so hot that "extensive measurements could not be made there on account of instrument contamination."[41] Stafford Warren later estimated that parts of it might have received 212–230 R.[42] The region was presumed to be uninhabited.

After the day's work, the cars the monitors were riding in themselves became radioactive. Several were abandoned temporarily. The monitors then showered long and hard and put on fresh clothes. John Magee confessed that he did not dare check himself after his long shower at South shelter. He simply dressed and left. His badge showed a dosage of 5.5.

The contingency evacuation plans for the region were not well conceived. In the first place, there was only loose agreement as to how hot an area needed to be before it should be evacuated. Several places around both Bingham and Vaughan showed readings of 6.5 R per hour, numbers which the official reports listed as "dangerously close to the evacuation limit."[43] At a July 10 meeting, Stafford Warren confessed that he would worry if the peak reached 10 R per hour.[44]

The scientists put the evacuation levels at a high figure. This was done in part for reasons of secrecy, but the physicians also recognized that there would be no untoward biological effects from such a one-time exposure. They were not much concerned with remote effects.[45] Awareness of the consequences of long-term, low-level exposure to radioactivity lay a generation in the future.

Some regions reached the high 15 R figure. Hot Canyon came in at 20 R. Even after the radioactive cloud had blown past, the canyon region still remained in the range of 3.5 R. Fortunately, the areas with the highest readings were all uninhabited.

The radiologists also concerned themselves with total dosage levels. Wright Langham remembered that the evacuation standards were set at more than 50 R over a period of a week.[46] Stafford Warren suggested that even 100 R in a two-week period would probably not be harmful, if the people involved received no further exposure. But he agreed to consider evacuation if the final total dose reached the 60–100 R range.[47] Accumulated dosage levels, however, would have been difficult to determine. The upper limits for evacuation always remained rather vague.

Second, if disaster had struck, Palmer and his soldiers could never have gotten all the people out. They did not know where everyone lived. The man ultimately responsible for the predicament was Groves, for when he fixed the day of July 16, he chose a period of unstable winds. Hubbard's predictions of probable wind directions for the cloud changed several times prior to the test, and this forced a limited number of military men to survey a vast area around Trinity: the northeast, the southeast, and the northwest. As a result, their survey remained incomplete. It was not until days later, as the monitors continued to check the area, that they discovered certain homes no one knew existed. The region of Hot Canyon turned out to be occupied after all.

Moreover, the 1940s proved very naive about radiation. For years after the test, people sneaked into the area to gather trinitite for souvenirs. The El Rio Motel in Socorro sold it over the counter to tourists. A Santa Fe bank gave away samples to new customers, with a warning "do not hold near body more than 24 hours." One woman even made a necklace from it. Eventually in 1946 the Los Alamos employees who kept trinitite in their homes were told that they had to sign a hazardous material receipt. While trinitite was not especially dangerous, they were told, improper handling of it over a long period of time could perhaps cause injury.[48]

When Helen and William Wyre of Bingham first spotted soldiers with little black boxes around their home on July 17, they were told they were checking for radioactivity. "We don't even have the radio on," said William Wyre. Several months later, Hempelmann cautioned Lt. Howard C. Bush that he probably was escorting too many parties into the restricted area. "I thought I was immune by that time," Bush replied. Bush later told a New Mexico reporter visiting the site not to worry about the radiation. "After all," he said, "my men and I live here and work here and it hasn't affected us."[49] General Groves told the senators in 1946 that if anyone were overexposed to radioactive materials, he simply took a vacation and in due time became all right again.[50] Such comments were common.

The scientists involved with the project also often became careless regarding protection measures. Most were more interested in their experiments than their exposure levels. Several even asked James Nolan how much radiation he would let them take.[51] The common understanding was that 10 R was maximum for a single day's exposure. Richard Watts calibrated his meters around that figure because he discovered that it was the upper limit most scientists were willing to absorb.[52]

The behavior of two army colonels brought in from Washington to view the test illustrated this rather naive 1940s attitude about radiation. Their comments at the time revealed they had little idea of the ramifications of the situation, and immediately after the explosion they returned to the Pentagon. They felt assured that all radioactivity that reached the ground was due to "rain out." Until the early 1950s, they publicly denied that any fallout resulted from the Trinity blast.[53]

CHAPTER SEVEN

The Aftermath II:
Cattle, Film, and People

In spite of the two colonels' denials, there had been a considerable amount of radioactive fallout after the Trinity test. The major part of northern New Mexico was covered with low activity. The coarser, heavier particles started to descend almost immediately, and when the cloud drifted over the Chupadera Mesa, the afternoon thunderstorms concentrated the fallout even further. Monitors eventually measured contamination in a path that extended from Ground Zero one hundred miles north to within fifty miles of Las Vegas, New Mexico, and northeast and north of Carrizozo; a narrow tongue east almost to Roswell (about a hundred miles); southwest to the Rio Grande and west to the Magdalena Mountain range. The original extent of the contaminated area has never been fully known. The one region of heaviest contamination, outside the restricted area, however, lay on the Chupadera Mesa, beginning about thirty miles north of Ground Zero and stretching north and northeast.[1] The mesa was a limestone outcrop with an average elevation around 7,000 feet, containing steep hills, piñon pine and juniper, and open, grassy meadows. It was also the principal grazing range of the valley.

The first to feel the effects from the radioactive fallout were the livestock that had been grazing nearby. Except for initial gamma radiation, the sheep seemed not to have been much affected. Only their faces lay exposed, and their wool at that time of year was about one and a half inches thick and very tight.

131

Moreover, wool is greasy, so any fallout particles deposited on it would have a difficult time penetrating to their skin.

It was a different story with the cattle. The first public evidence of the fallout came from the cattle that had been grazing near Trinity Site. About three to four weeks after explosion, several cows began losing hair. The hair quickly grew back, but instead of being normal Hereford color, it came in white. Hair on other cattle in the area began to show similar white mottlings, or streaks. No two were marked alike. Some had white backs, others, white sides or miscellaneous blotches. A few were covered with white so thoroughly that it looked as if they had been dusted with frost.

Cattle from the Ted R. Coker ranch, ten miles from Bingham, New Mexico, and about thirty-five miles from the site, and at the adjoining H. O. Bursum and C. M. Harvey ranches were all rather heavily affected. The cattle all appeared healthy, but because of the strange color markings, the purity of the pedigree was questioned, and the ranchers had to take a cut in their price.[2]

Unpainted fence posts on some ranches turned white, as did half the beard of Bingham rancher Will Wyre. A black cat belonging to one of the hands at the Bursum ranch also turned half white. A passing tourist took a fancy to the cat and bought it for a dollar. Unfortunately for science, but probably fortunately for the cat, it escaped further scrutiny. At first the ranchers were puzzled. The cattle nearest the blast were not as affected as those farther away, so the cause was not immediately clear. Some attributed the change to high winds or dry weather or a strange new disease. Holm Bursum felt that he had probably purchased a bad bull. In early October, a Carrizozo attorney, representing the Red Canyon Sheep Company, filed a claim against the government for injury to his client's animals.[3]

So, in November of 1945, Louis Hempelmann of the Los Alamos Health Division and Lt. Colonel Stanley L. Stewart, who had grown up on a Colorado ranch, made several trips to the Trinity region to inspect the cattle. They bought four cows and transferred them to the Hill to determine if the blast really was the cause of their discoloration.

Suddenly, the cattle became famous. New Mexico State Cattle Sanitary Board inspectors posed proudly for photos next to their "atomic calves."[4] The "gray overnight" cattle or "rada cows" were exhibited in El Paso and Alamogordo. One entrepreneur made plans to open an atomic bomb museum in Albuquerque, where the cows could go on permanent display.[5]

It was not long before Los Alamos tests confirmed that the discoloration was, indeed, from Trinity fallout. In December 1945, Hempelmann and Stewart returned to the area with instructions to survey the damage and buy up all cattle which could not be sold as prime beef. They were to pay full market price to the ranchers. Perhaps 350 cattle were affected to some degree by the discoloration, but only about 75 were badly damaged in terms of market value. Bursum still had 200 left with white splotches after he sold the most marked to the government.

Using Manhattan Engineer District funds, Colonel Stewart bought the cows. The seventeen most badly affected were shipped to Los Alamos while the remainder were put on trains for Oak Ridge, Tennessee.[6] It was not until the cattle were herded into one corral for shipment that one could see the extent of the damage. Some had healed and sported splotches of white. (This reminded ranchers of how a severe saddle sore on any color horse will always grow back in white.) Other cows still showed scabs, which rubbed off as they brushed together. It looked as though they had all been scalded by something.[7] Damaged cattle are not allowed to cross state lines, of course, so these cows had to procure a special sticker which said, in effect: These cattle are not diseased. They have been burned by the atomic bomb.

Although government spokesmen were officially rather coy about their purchases—one representative from Los Alamos declared that the gray spots were simply due to age[8]—privately they had determined to study the cattle thoroughly. Complaints were made, however, that no specific genetic studies were possible because the vital facts of age, ancestry, and dosage each cow received were not available.

The cattle were given red carpet treatment, but little happened. While they retained their markings, no further graying emerged. None died of unexplained causes. They reproduced normally. Seven calves of the affected cattle were also kept at Los Alamos, but no one could detect any genetic changes or mutations.[9]

Eventually the cattle became somewhat of a nuisance. In 1947, Hempelmann suggested that they could best be quartered in the

Los Alamos medical director's office.[10] This admirable recommendation seems not to have been followed, however, for they were contracted out to Dr. Thomas L. Shipman, also of the Health Division, and to other area ranchers. Periodically they were examined by the Los Alamos veterinarian, Dr. R. E. Tompsett. Finally, in the late 1940s and early 1950s, the cattle were all shipped to Oak Ridge. Not everyone was happy with this decision, however. "I certainly had my mouth all set for a new piece of barbecued beef," complained Thomas L. Shipman, "and it now looks as though that pleasure will be enjoyed in Tennessee."[11]

In 1949, the Atomic Energy Commission released some figures on the cattle. After extensive tests of seventy-three head that were most heavily exposed, they announced that the cattle appeared normal in every way. Of fifty cows, forty-nine were bred successfully. One exposed bull was bred to exposed cows, and the thirty-three calves were completely normal.[12] Some of the cattle were slaughtered and eaten. The radiation levels of the meat showed only slightly higher than normal background radiation.[13] Several of the cattle were turned over to the Animal Husbandry and Science Department of the University of Tennessee, and a graduate student, James B. Bird, wrote a master's thesis on their condition. In 1952, he concluded that "no gross differences were found in the performance of the irradiated cattle and control cattle nor in their offspring."[14]

While no genetic changes emerged, at the end of their lives several of the more severely exposed cattle did develop skin cancers on their backs. Their chief discomfort, observed Hempelmann, was that they suffered from "mild beta burns."[15]

Another example of the spread of radioactive fallout came from the contamination of several batches of film manufactured by Eastman Kodak Corporation of Rochester, New York. The discovery of this reads like a detective story.

In the fall of 1945, several batches of sensitive industrial X-ray film were found to be flecked with imperfections. The damaged areas ranged from ten to several hundred spots on fourteen-by-seventeen-inch sheets of X-ray film. The job of discovering the cause of the contamination was given to Henry Clyde Carleton. Carleton was an expert in tracing film imperfections. It was he, for example, who had earlier discovered that if a Kodak assembly line employee had put iodine on a cut finger, traces of it would show up on the films for as long as two weeks.

Carleton and his staff began checking. They examined the base of the film. They checked for impurities on the photographic emulsion, and in the emulsion itself. Nothing. Finally, by accident, someone discovered that the spots on the film lined up perfectly with spots on the strawboard dividers separating the films in their cartons. Here was the answer. The strawboard had fallout particles enmeshed in its fibers.[16]

Dealing with radioactive contamination was not new for Kodak. Because of earlier problems with radium-contaminated cardboard from eastern manufacturing plants, Kodak had switched its source of supply to the Midwest. They relied heavily on two paper mills, one at Vincennes, Indiana, located on the Wabash River, and the other at Tama, Iowa, on the Iowa River. The mills lay about five hundred miles from each other and relied upon different watersheds.[17] The Vincennes strawboard was produced on August 6, 1945, and the Tama strawboard in September.

Extensive tests by the Kodak Research Laboratories, working in conjunction with the University of Rochester, confirmed that the contaminated material contained the radioactive isotope cerium 141, one of many fission products produced by the detonation.

Although the scientists were not positive, they felt it likely that the straw itself was not contaminated. Instead, the trouble lay in the river water used in the manufacturing process. Late summer rains washed the radioactive particles from the soil surface into the river. The radioactive materials involved, of course, were the size of clay and silt. Only a series of events— the concentration of fallout particles in the water, the filtering involved in the paper pulp manufacturing process, and their use in packing sensitive X-ray film—allowed the contamination to be detected.[18]

In late May of 1946, Eastman Kodak made its findings public. The two mills were over one thousand miles from New Mexico. Since other readings discovered unexplained radioactivity in Maryland, the *New York Times* concluded that "The single bomb exploded in New Mexico contaminated the air over an area as large as Australia."[19]

From July 1945 to early 1947, scientists from Los Alamos Laboratory made periodic, albeit infrequent, visits to the Trinity region to inquire about the livestock, land, and residents. Pressures were such, however, that no systematic study was done until the summer of 1947.

Initially, the Atomic Energy Commission was reluctant to fund any studies of the Trinity backcountry. The war had ended, and considerable apathy prevailed. Moreover, the AEC lawyers feared that investigators might find something that would induce lawsuits. Stafford Warren, however, now dean of the UCLA Medical School, insisted that the region be scientifically explored.[20] Eventually the Division of Biology and Medicine of the AEC gave him permission and funds to conduct the first ecological study of the Trinity area. Continuing follow-up studies occurred in August of 1948, 1949, 1950, and 1955.

The UCLA group included ornithologists, herpetologists, entomologists, biologists, soil scientists, health physicians, and botanists. When they visited the site in 1947, a detachment of MPs still guarded the area. An eight-foot-high guard fence, about 1,600 feet in diameter, surrounded Ground Zero and kept out everything except plants, birds, and small animals. The fence remains today, but the MPs all left in the summer of 1948.

From 1947 to 1949, the UCLA studies were secret. Security was such that Los Alamos could not tell the wife of a local Standard Oil station owner that she should probably dispose of the shoe box of trinitite that she kept in her closet. (She eventually did so on her own.) Local sentiment was not especially favorable toward "federal people," and the investigators occasionally could hear ricocheting rifle shots in their vicinity. It took the efforts of understanding ranchers, such as Ted Coker, to ensure tranquility.[21]

Although it had been over two years since the Trinity detonation, the survey group found effects of the blast and heat still quite visible. The land under Ground Zero was depressed about six feet, like a saucer, the edge of which sloped away from the center. It seemed as if a gigantic sledge hammer had hit it. This zone covered a radius of about 250 feet. All vegetation to about 2,000 feet from Zero had been destroyed by the blast, but by 1947, it had begun to come back. Deep-rooted plants, such as yucca, began growing rather easily. One Larrea (a perennial shrub) was found within the inner 300 feet. It was so large that one observer claimed it "must have germinated after almost the first rain after the bomb was set off."[22]

Surrounding the crater lay a zone of fused surface soil, the trinitite, often in nearly solid sheets. Ranging from 3/16 to 1/2 an inch thick, it covered approximately seventy-three acres. It also crumbled very easily. Beyond the trinitite lay a zone of almost no plant life, and outside the fence the grasses were slowly returning. Preliminary studies indicated that the radioactivity had not reached the root zones of most of the plants. The deep neutron penetration under Ground Zero had interacted with the elements of the earth. Thus, it was also unavailable to plant life at this time.

The scientists noted some rather disturbing things in their 1947 summer survey. They discovered plutonium in the soil and on plants at numerous locations for a distance of eighty-five miles outside the fenced-in area. A later survey estimated that another area of maximum concentration had been deposited on the Chupadera Mesa.[23]

While the deep-rooted plants had not absorbed much activity, preliminary findings did indicate that the shallow-rooted grasses on the mesa had trace amounts of radioactivity. Cattle feces gathered there all showed marked radioactivity, for the cows ingested both plants and soil as they grazed. The scientists concluded that the area where radionuclides might be available to plants, and hence to man, was larger than had been expected.[24]

The ornithologists discovered a number of what they assumed to be radiation-damaged birds. Five horned larks had malformations of their feet and claws. One three-month-old jay had a desiccated foot.[25] The biologists reported a significant number of rodents with eye cataracts, in a ratio far above the normal population.[26] Several ladybird beetles also had abnormal spotting on their backs. While the biologists looked for major genetic changes, no one discovered any. Various tentative suggestions were made to explain their absence, but no firm conclusions were drawn. The questions that sparked the 1947 study remained unanswered as to the health of the region: Were the springs likely to become contaminated? Could migrating animals spread radioactive contamination? Should grazing restrictions be imposed? Should some families be removed from their ranches for five to ten years, or even longer, because of low-grade exposure?[27]

The 1947 survey team drew no conclusions on these matters. Instead, it urged that additional studies be conducted the next year.

The 1948 study reached more optimistic conclusions, and all the later studies supported this one. No damaged birds and rodents were identified. From this they reasoned that the damaged animals found the previous year had been affected by the initial detonation, but that the effects were not cumulative. Numerous rodents were trapped within the fenced-in area and on the mesa. While some of these small animals showed small amounts of radioactivity in their bodies, it was limited almost exclusively to their gastrointestinal tracts. This meant that while they were eating radioactive materials, none of the radionuclides were being absorbed by their bodies. Instead, they were passing through.[28] Of 402 small animals collected during the 1947–48 surveys, only 38 showed significant count. All of these were only twenty times the normal instrument background level.

The 1948 survey team also found that the cattle on the mesa were not accumulating significant amounts of radioactive materials in their systems. That summer the scientists purchased eight head of cattle from Ted Coker; all of them had been grazing in areas that measured significant levels of radiation. These cows still had the white spots on their backs from the fallout. In November the cattle were trucked to Schwartzman Packing Company in Albuquerque, slaughtered, and the meat carefully examined by the chief city meat inspector. He deemed it acceptable for consumption. The carcasses were then shipped to UCLA for further radio assay. The scientists, however, could find no appreciable difference in radiation between them and other cattle taken from the Los Angeles area. In fact, the cattle from the Los Angeles feed lots registered slightly higher than those from the mesa.[29]

The scientists discovered that radioactive material seemed to be spreading.[30] In spite of occasional flash floods, the main factor in distributing the material proved to be the winds. This was brought home abruptly to the crew when the 1950 survey team sat through five heavy dust storms within a month. "The entire valley—some 3,000,000 acres—is on the move," complained laboratory chief Albert W. Bellamy.[31]

The scientists also did numerous experiments with trinitite.

In one test, they pulverized it for six weeks to turn it into very fine powder. This was then mixed with other soil and planted. The powdered trinitite was also fed to rats in daily, measured amounts. Gradually, they found that trinitite did not behave like the water-soluble radium that had so affected the dial painters. Instead, the material acted as if it were shielded in glass. Trinitite did not easily enter the life cycle of either plants or animals.[32]

While the Jornada is not a true desert, it still receives only six to ten inches of rainfall a year. In addition, it has a very high level of evaporation. This all reduced the minute solubility of the fission products and tended to confine them to the top two inches of soil, a fact which proved true even nine years after the detonation.[33] Since little or no leaching occurred, this meant that most of the fission products remained locked in the glasslike fallout particles that lay in the soil above the root level of the area's plants. As nearly as they could tell, the fission products were not being absorbed by the plant life. The 1948 study concluded that there was no significant difference in activity between crop plants from contaminated agriculture areas and those from uncontaminated areas.[34]

The plants in the mesa continued to show low-to-moderate activity, however, but the scientists felt this could be traced to the fission products dropped by the original Trinity cloud. No amount of washing could remove every trace of radioactivity from the plants, because of the resinous surface. Local rain showers after the blast had helped wash out these radioactive particles in certain areas, but the moist bark and leaves proved able to hold the radioactivity especially well.[35]

The only area where the radioactivity penetrated beyond the top level of soil was within the fenced-in region, especially in the crater itself. Immediately under the center of the remaining stubs of the 100-foot steel tower, the neutrons penetrated deep into the soil. In 1947, activity was found at a depth of forty-two inches. The neutrons would probably continue to penetrate until they ran out of energy.[36] This area covered approximately 145 acres.

The 1948 report concluded optimistically that "a human being would have to go to some trouble to expose himself to the minimum permissible daily dose of beta-gamma irradiation." So, in 1949, the government ended the secrecy regarding the various studies of the Trinity region. As Air Force Major D. M. Brown told a group of local New Mexico citizens: "Our data indicates definitely that, except for the fenced in crater, no hazard from external exposure to beta-gamma radiation exists either to man or his domestic animals." Yet to be certain, the government promised the local ranchers that they would return each year. They also noted that "many interesting problems of the redistribution of radioactive materials in soil, plant, and animal life cycles are under study."[37]

While this was, perhaps, true, the continued studies—classified until the early 1960s—told a more cautious story. Although couched in bland jargonese, the message remained clear. The 1948 study noted that "it would be rash to conclude, in the absence of specific information, that now no hazards associated with products of the bomb detonation exist in this area, the harmful effects of which may not appear for a number of years."[38] As the 1950 study phrased it:

The biological significance of the data presented in this report cannot be evaluated at this time. It is not possible under the present circumstances of the Field Study to assess the potential hazard of the plutonium found to the people and cattle living on the Chupadera Mesa.

Two pages later it noted:

In the absence of better information, it would seem logical to suspect that conditions hazardous to man are not absent from the areas on the Chupadera Mesa, particularly if occupancy occurred over a considerable number of years.[39]

All the surveys urged that studies of the area be continued. The long half-life of the materials involved—strontium 90, twenty-eight years; cesium 137, twenty-nine years; plutonium, 24,000 years—meant that all conclusions reached had to be tentative. Continued monitoring was deemed necessary.

This, however, was not to be. The 1955 survey was the last for many years, and part of that had to be funded through a special grant from the biophysics department of the UCLA Medical School. AEC officials were unwilling to invest any more money in Trinity. The head of the Division of Biology and Medicine was never sympathetic to field studies, and more pressing tests in the Pacific and in Nevada claimed the AEC's attention. Several of the Trinity Site investigators moved to Nevada and tried there to solve the questions first raised by the Trinity explosion.

Land, plants, and even animals lend themselves easily to scientific study, but people do not. In 1945, of course, no scientist could have known about the long-term consequences of low-level radiation. Still, however, the government showed considerable lack of foresight in dealing with some of the area residents.

The early studies of the people in the fallout region were casual. On July 16, monitors roamed through the area striking up conversation with the various local residents. They examined the two broken windows at R. H. Dean's Bingham General Store and chatted with H. A. McSmith, proprietor of the nearby White Store. One monitor had his car serviced at 6:15 A.M. on July 16 at John Muncey's Station in Carthage. Postmaster and a native of Carthage since 1902, Muncey (who had slept through the blast) complained that Carthage was too quiet. The only noises he ever heard, he grumbled, were those of the desert animals.[40] Everyone seemed normal, so nothing was done.

On July 17, Lt. Colonel Hymer Friedell and Louis Hempelmann investigated the Hot Canyon region. Much to their dismay, they discovered a family living less than a mile from there. It consisted of Mr. and Mrs. Raitliffe, both of whom were over fifty, and their ten-year-old grandson. The house was not registered on any of the monitors' maps, and no one knew it was there. A second ranch, also unknown to the army, lay nearby.[41]

Two days later, the physicians and monitors met in Los Alamos to analyze the results of their efforts. Their figures showed that the radioactive decay was considerably more rapid than anyone had anticipated. Although they discussed the question of evacuating the Raitliffe family, they decided that it was not necessary at that time.

On August 12, Stafford Warren Louis Hempelmann, and an army captain made an extensive return field trip to the area. Warren soon returned to the Base Camp, but the others visited Bingham, White Store, and Hot Canyon. They took measurements and chatted with the people. The Dean family in Bingham and the McSmith family in White Store both seemed equally healthy. Afterwards, the men explored Hot Canyon. They visited the two-room adobe house of the Raitliffes and had an extensive conversation with them.

The elder Raitliffes had slept through the detonation, and the grandson had ridden his horse to Bingham early that morning. Since he did not return until late that night, he probably missed most of the heavy exposure that fell on Hot Canyon. Everyone appeared healthy except Mr. Raitliffe, who complained of "nervousness, tightness in the chest, and poor teeth." These symptoms had bothered him for years. The observers also noted that the Raitliffes' tin roof was used to collect rainwater for drinking. Since it rained the night after the shot, this must have washed the radioactivity into their cistern. Several measurements were taken in the area, ranging from .004 R per hour to .027 at the highest.[42]

The fallout, of course, had blanketed the region. Stafford Warren estimated that a stretch of Highway 380 south of Bingham received about 50 R total.[43] A ranch house in Bingham was calculated to have received an initial dose of 7 R per hour and the inhabitants to have absorbed 40–50 R within two weeks; they might have received as much as 60 R within four weeks.[44]

The nearby ranch families were all exposed to radioactivity. Ted Coker actually stood under the cloud for some time and had the particles fall on him. "It smelled funny," he told the scientists later.[45] Another family, which lived about four miles in from Coker, had a fourteen-year-old daughter, who was out playing in the yard when the cloud moved over. Yet no efforts were made to follow up on the health of these people.

Radiation physics professor Ernest J. Sternglass has argued that the Trinity fallout caused an increase in infant mortality for a generation of children in Alabama, Texas, Arkansas, Louisiana, Georgia, the Carolinas, and Mississippi. This is unlikely, however, for the radioactive cloud did not pass over those states.[46] The National Association of Atomic Veterans, an organization devoted to aiding soldiers who may have radiation-related ailments, has a Trinity coordinator. His job is to assist those soldiers who claim that their health was adversely affected by the Trinity test.[47] In 1983, the Defense Nuclear Agency published their findings regarding exposure levels of all personnel at Trinity.[48] At this writing, no Trinity-related radiation case has been sustained by the government.

Whether the Trinity fallout had any effect on the deaths by cancer of Ted Coker, Mrs. Helen Bursum, or the numerous members of the Dora Chavez family, who lived twenty miles away in the Capitan Mountains, cannot now be decided. It is a question that is impossible to resolve because sufficient records were never kept. Yet, as far as can be determined, no major wave of illness swept over this sparsely settled area to strike down its inhabitants.

One should not discount the elaborate preparations the army, the Los Alamos scientists, and the physicians made to ensure the safety of the area. The health physics arrangements by the medical group were the most sophisticated that the world had ever seen. In fact, the precautions taken at Trinity were more elaborate than those at several subsequent Nevada tests. But because the scientists stood at the edge of knowledge, they were not certain of what might occur. As it happened, a series of fortunate circumstances came together to ensure that the Trinity experiment did not turn into a disaster for New Mexico.

First, the yield from the blast proved much larger than expected, and the meteorological conditions turned out to be perfect to allow the rapid rise of the cloud. So, it quickly rose not to the anticipated 12,000 feet but to 40,000 feet. Consequently, it spread radioactive material over a wider (and hence safer) area. Had it reached only the 12,000-foot level, fallout intensities would have increased dramatically. Hot Canyon, for example, might have reached 40 R per hour.

The particles of soil the fireball picked up and made radioactive were smaller than had been expected. Many of them condensed out in the form of slowly drifting fine clay, which was mixed with water vapor. The initial calculations of Hirschfelder and Magee proved basically correct—about 49 percent of the radioactive material was deposited in a few hours and 50 percent over several months—but this occurred over a much larger area than anticipated.

Moreover, the explosion occurred under conditions that had been thought likely to induce thunderstorms, yet none occurred. Since the spotty afternoon rain later concentrated the radioactive materials onto various "hot spots" on the mesa, this proved fortunate, indeed. A gigantic storm in the wake of the blast would have been disastrous for everybody.

Finally, plutonium and the other fission products contained in trinitite and other silicate fallout material did not behave like radium. When ingested, the radioactive material did not become available to bones and other tissues. Instead, most of the fallout materials were eliminated by the animals in the region. The radionuclides also seemed to stay in the surface two inches of soil. The high evaporation and low rainfall of the area meant that the deep roots of the plants would absorb only a small amount. Consequently, except for the crater area, where the zone of intense radioactivity was decreasing steadily with the years, the radionuclides in the radioactive deposits from Trinity never became part of the ecological system of the region; at least that was what a decade of study concluded. None of this had been planned, however; it just happened.

Five days after the Trinity test, Stafford Warren wrote a long letter to General Groves. In it he warned that the Jornada del Muerto region was too populated for further nuclear explosions. Instead, he urged that any future test be held at a site with a radius of at least 150 miles without people.[49] It was not until the mid-1960s that Warren confessed he had finally stopped worrying about the Trinity explosion. As Louis Hempelmann observed in a 1982 interview, "We were just damn lucky."[50]

CHAPTER EIGHT

The International Legacy

The most momentous consequences of Trinity were political ones. The explosion proved that the scientists' theories were correct. As such, it set the stage for both the immediate end of World War II and the nuclear arms race.

Groves quickly flashed word to George Harrison, acting chairman of the Interim Committee on S-1 (the bomb), in Washington. Harrison, in turn, cabled the news to Secretary of War Henry L. Stimson in Potsdam. The message arrived there at 7:30 P.M. on July 16. It noted, "Operated on this morning. Diagnosis not yet complete but results seem satisfactory and already exceed expectations. Local press release necessary as interest extends a great distance. Dr. Groves pleased. He returns [to Washington] tomorrow. I will keep you posted." The next evening another message from Harrison stated: "Doctor Groves has just returned most enthusiastic and confident that the little boy is as husky as his big brother. The light in his eyes discernible from here to Highhold and I could hear his screams from here to my farm." This translated as, had the blast gone off in Washington the sound would have traveled to Upperville, Virginia, about forty miles away, and the light could be seen at Highhold, Stimson's Long Island farm, over two hundred miles away. After he read the cable, Stimson turned and said to his companions, John T. McCloy and Harvey H. Bundy, "Well, I have been responsible for spending two billions of dollars on this atomic venture. Now that it is successful I shall not be sent to prison in Fort Leavenworth."[1]

A detailed report was to follow shortly, but Stimson immediately told Truman and Secretary of State James Byrnes. Both were delighted. "The President was tremendously pepped up by it," Stimson recorded in his diary when he first told Truman, "and spoke to me of it again and again, when I saw him. He said it gave him an entirely new feeling of confidence. . . ."[2] Churchill later remarked how forceful Truman became in the next day's negotiations because of this new piece of knowledge.[3] On July 18, General George C. Marshall told the Combined Chiefs of Staff, who were still planning on an invasion of the Japanese home islands, of the successful test. Groves and Farrell worked two days and nights on their complete report, and it arrived, by special courier, on the morning of July 21, just before noon. The report was very detailed and contained several additional supporting documents.[4]

Should Truman have offered a full disclosure of the Trinity success, with additional information as required, to Joseph Stalin at Potsdam? Some scientists, most notably Leo Szilard and Niels Bohr, had been urging such international cooperation for years. They argued that the Russian scientists were bound to discover nuclear fission before long and that an open sharing by the Americans would obviate any race for armaments. In fact, Oppenheimer later claimed that Bohr was in the United States primarily to lobby for this position.[5]

Neither Szilard nor Bohr, however, made much headway. Szilard was unable to gain an audience with President Roosevelt and had a most unproductive discussion with James Byrnes. Bohr's scheme infuriated Winston Churchill. In September 1944, the prime minister and Roosevelt instructed the FBI to tail Bohr out of fear that he might try to share atomic secrets with the Russians on his own.[6] Henry L. Stimson later urged international cooperation in a dramatic September 1945 cabinet meeting, but this always remained a minority position.[7]

As a sort of compromise measure, Truman agreed to tell Stalin of the success at Trinity, but with the secret hope that he would not press for exact details. The occasion presented itself on July 24 when the president walked over to his Russian counterpart and casually remarked that the United States had just tested a new and powerful weapon. Stalin said he was glad to hear it and said he hoped America would make "good use of it against the

Japanese."[8] After this rather awkward exchange, Truman withdrew, and nothing more was said of the matter.

In his recent memoirs, Marshall George Zhukov claimed that Stalin understood perfectly that Truman had been referring to the atomic bomb.[9] Truman, however, always felt that Stalin did not understand its significance.[10]

A February 1983 *Isvestiia* article by Soviet scientist Anatoli P. Aleksandrov offered a glimpse of the World War II Soviet atomic program. Aleksandrov reported that Soviet nuclear research first began in the early 1930s and that by 1941 the USSR had one cyclotron in operation and two more under construction. Even the invasion of the German Army did not disrupt the program, for all Soviet facilities and personnel were moved into the interior at Kazan. They continued their work there until 1945.

By the spring of 1942, the Soviet scientists realized that all publications on nuclear physics had vanished from the western scholarly journals. The names of the scientists who had been working in the field also disappeared from the scientific literature. When Stalin was informed of this, he summoned physicist Igor V. Kurchatov to Moscow and instructed him to begin work in earnest. The Soviets had concluded that both Germany and the United States had secret atomic weapons programs under way. From the onset, Kurchatov expressed more fear over the intentions of the American ally than of the German enemy. After the war, Soviet nuclear efforts continued steadily, and on August 29, 1949—three to five years ahead of western forecasts—they exploded their first atomic bomb.[11]

Although Aleksandrov showed that the Soviets had their nuclear program well under way by the Potsdam Conference, Truman was probably correct that Stalin had no real conception of the power of a nuclear explosion. A World War II "blockbuster" bomb was equivalent to ten tons of TNT. In the Trinity betting pool, many Los Alamos scientists guessed that the bomb would be only eight to ten times as powerful. Had this proven correct, atomic weapons might not have made economic sense. It was not until *after* the Trinity test—with an explosion close to 20,000 tons of TNT—that the Los Alamos scientists fully comprehended the destructive potential of a nuclear blast.

Moreover, Truman approached Stalin on Tuesday, July 24. This was too early for any Soviet report on the Trinity explosion to have reached him. By the time of the Potsdam Conference, the Soviets knew only that the American scientists were preparing to test the proposed nuclear weapon.

The only Soviet agent at Los Alamos with real scientific training was physicist Klaus Fuchs. In May 1945, Fuchs had informed his courier Henry Gold that an atomic test was scheduled for sometime in July.[12] There is no evidence that this information was acted upon, however, and it is doubtful if any Russian agent witnessed the experiment. While Fuchs observed the Trinity explosion firsthand, he was unable to relay information about it. It was not until September 19, 1945, that Fuchs told Gold of the sense of awe that he and the other scientists had felt when they observed the blast.[13] Thus, it is highly likely that when the bombs were used in combat three weeks after Potsdam, the Soviet Union was caught by surprise. In November 1945, W. Averell Harriman reported from Moscow that the Soviets had already accused the Americans of trying to intimidate them with atomic weapons.[14] The arms race had begun.

What if Truman had used the Potsdam Conference to offer to share the knowledge of the American and British discoveries? What if the Americans had put forth some immediate plan for the international control of atomic weapons? Could the arms race have been avoided? At this juncture, no one can say for certain, but a glimpse of the road not taken is always tantalizing.

It might well be argued, however, that in 1945 such cooperation was not possible. The United Nations still lay on the drawing board, and there existed no political machinery adequate for such international control.[15] Twenty-four years of intense distrust between the Soviet Union and the United States proved hard to overcome, even after four years of cooperation against the Axis powers. The Soviet nuclear program was even more secret than the Americans', and the Kremlin leaders all harbored grave suspicions about the United States. Moreover, there was little in Stalin's personality that suggested the possibility for close cooperation with Western leaders. In fact, he later severely punished those who advocated a more lenient approach to the West. Whatever conclusion a person draws, however, one important milestone on the road to the present Soviet-American arms race must be that unproductive encounter between Truman and Stalin on July 24, 1945.

The second major political consequence of the successful Trinity detonation was the decision to utilize atomic weapons in combat. According to William L. Laurence, shortly after the explosion, the following exchange occurred between Generals Farrell and Groves. "The war is over," said Farrell. "Yes," replied Groves. "It is over as soon as we drop one or two on Japan."[16] Laurence later termed the Trinity shot "a full dress rehearsal" for the bomb's use on the Japanese home islands.[17]

Led by Truman, the Allies drafted a statement demanding that Japan surrender immediately. This Potsdam Declaration did not mention the atomic bomb but threatened Japan with "complete and utter destruction" if she continued the war. The message was delivered through diplomatic channels and also dropped in leaflet form over the home islands. The Japanese government replied with the ambiguous word *Mokusatsu.* The United States interpreted this (perhaps incorrectly) as "to ignore."[18] The Americans responded on August 6 by dropping the first atomic bomb on Hiroshima. The Soviet Union entered the war against Japan on August 8, and the following day America dropped a second bomb on Nagasaki. On August 10, Japan surrendered.[19] The terrifying consequences of these bombs, however, unleashed an avalanche of literature that asked if one or both were necessary. This question is not likely to be resolved in the near future.[20]

There are two basic positions on this question. The official argument maintains that the dropping of the bombs ended the war abruptly, that it enabled the Japanese to surrender without losing face. The Joint Chiefs of Staff always had great respect for the Japanese military. General George C. Marshall estimated that if the Allies had to invade the Japanese Islands, the potential losses would run from perhaps a half a million to a million men. In early August 1945, General John Dudley went to Manila to help the Sixth Army plan a landing on the main Japanese island. He was annoyed to find that his outfit was scheduled only as a second echelon. He changed his mind, however, when he discovered the reason. The second echelon was given the same task as the first. "It was clear to me then," Dudley recalled in a 1980 Los Alamos colloquium. "They expected the first echelon to be wiped out in the invasion. The second echelon would get the thing done."[21]

The shock of the bombs allowed the Japanese government to

surrender immediately. It let them end the war with honor. Even
with the bomb, the peace wing of the Japanese government barely
squelched an attempted coup by the militarists. Terrible though
the destruction was, in the long run it saved both Japanese and
American lives.

This point of view may be found in most contemporary comment immediately after the conflict, in the recently declassified
papers on the Manhattan Project, in the memoirs of Harry Truman, and in numerous books and articles.[22] It is especially well
stated in the works by State Department historian Herbert Feis
and Admiral Samuel Eliot Morison. The atomic bomb, concluded
Morison, "was the keystone of a very fragile arch."[23]

Every anniversary of Hiroshima brings renewed speculation.
Writing in an August 1983 issue of the *Journal of the American
Medical Association,* Japanese physician Taro Takemi agreed with
Feis and Morison. The only Tokyo doctor studying nuclear physics in 1945, Takemi was called to Hiroshima to investigate the
explosion. Using a crude method of analysis by radioscope, by
5:00 P.M. on August 8, he determined that the weapon had, indeed, been an atomic bomb. He informed Japan's Count Makino,
who immediately sought an audience with the Emperor. After
he heard the news, the Emperor agreed to accept the terms of
the Potsdam Declaration. "When one considers the possibility
that the Japanese military would have sacrificed the entire nation if it were not for the atomic bomb attack," Takemi wrote,
"then this bomb might be described as having saved Japan."[24]

This position was graphically illustrated in an anecdote told
by Arthur H. Compton. When Compton visited Japan after the
war, the Japanese reporters always asked him why America used
the bomb. Once Compton responded: "Would you have preferred
that we should have let the war run its natural course without
using the bomb?" In that case, the reporter admitted, he probably
would not have been there to ask his question.[25] Some form of
this theory is probably held by the majority of the American
public, certainly among those who were politically aware during
World War II.

The other position, now termed "revisionist," is less sanguine.
Germany surrendered May 8, 1945, without discovering the se-
cret of atomic weapons. Japan was on the verge of collapse. By
July of 1945, American officials knew that she had already ap-

proached Russia about possible mediation. There existed little immediate need to drop the bombs. Instead, argue the proponents of this position, America dropped the atomic weapons primarily to intimidate the Soviet Union, both in Asia and in Eastern Europe. The dropping of the bomb, concluded British Nobel Prize winner P. M. S. Blackett, was "not so much the last military act of the Second World War as the first major operation of the cold diplomatic war with Russia."[26] Numerous other spokesmen have agreed.[27]

Several top-level American military leaders have also argued against the necessity of using atomic weapons against Japan. Their statements have (perhaps unwillingly) fueled the revisionist position. General Dwight Eisenhower did not want the United States to use the bomb, and Admiral Chester Nimitz always maintained there was no need for it.[28] Admiral William D. Leahy argued that "the use of this barbarous weapon at Hiroshima and Nagasaki was of no material assistance in our war against Japan. The Japanese were already defeated and ready to surrender because of the effective sea blockade and the successful bombing with conventional weapons."[29] The postwar, official U.S. Strategic Bombing Survey also concluded that the atomic bombs did little to defeat the Japanese. They were ready to surrender at any moment. General George C. Marshall admitted only that the bombs shortened the war "by months."[30] General Curtis LeMay, chief of staff of the U.S. Strategic Air Forces in the Pacific, maintained the war would have been over "in two weeks," both without the Russian intervention and without the atomic bombs. He credited the sustained Air Force attacks with rendering the Japanese defenseless. "The Atomic bomb," he said, "had nothing to do with the end of the war at all."[31]

From the early 1960s onward, this view has steadily gained ground. Its most able proponents have been historians Gabriel Kolko, Gar Alperovitz, and William Appleman Williams.[32] Recently, historians Barton J. Bernstein and Martin J. Sherwin have refined these views somewhat, but both acknowledge that concern over intimidation of the Russians played *some* role in the decision to drop the two weapons.[33] This interpretation grew in popularity during the antigovernmental period of the 1970s, and some version of it probably holds sway in most academic circles today. "I do not myself see how the evidence can be read in any other way," said noted sociologist Kai Erikson.[34]

There is, however, another perspective from which to view the events that flowed from the success of Trinity: the simple question of the momentum of the Manhattan Project. As Martin J. Sherwin has correctly pointed out, "What emerges most clearly from a close examination of wartime formulation of atomic-energy policy is the conclusion that policy makers never seriously questioned the assumption that the atomic bomb should be used against Germany or Japan."[35] As a result, it became almost a foregone conclusion that some action would follow immediately after the bomb's completion. As J. Robert Oppenheimer later put it, "the decision was implicit in the project."[36]

Because Harry Truman always assumed full responsibility for ordering the dropping of the bomb, historians have been misled about his "decision" to do so. The facts suggest that Truman did not really make any "decision." Instead, the momentum of the Manhattan Project made the decision for him. As perhaps the most complex human undertaking in the history of the world, the project demanded that all things be set in motion years *before* they would be ready. Alexander Sachs recalled that in 1939 he explained this concept to President Roosevelt by means of an analogy from music. A composer of, say, a round, had "layered" his music and in a sense "telescoped" the time involved to sing it. Because of time pressures, Sachs urged that the atomic bomb be created in a similar fashion. "When you start one part of the project," he advised the president, "assume you have finished it successfully, and start the next as if you had."[37]

At the moment the army assumed control, General Groves gave the Los Alamos scientists their mission. They were to test a bomb as soon as they had sufficient fissionable material.[38] The production of the nuclear material always determined the time scale of the project. As there would never be a sufficient amount of U-235 from Oak Ridge, the scientists had no plans to field test the uranium bomb. Only the fact that plutonium production was faster permitted the Trinity test; in fact, a third plutonium bomb was ready for use on August 24, two weeks after Japan's surrender.

During the early days of the war, when the atomic race was at its height, few people thought of alternatives other than military use. In 1946, physicist John Simpson admitted that during the early days of the project the scientists at Los Alamos spent little time in thinking about the possible effects of what they were trying to make. Alexander Sachs alone recalled that in December 1944, President Roosevelt had given a general acquiescence to Sachs's plan for a test explosion (presumably at Trinity) to be observed by all concerned nations and members of all the world's great religious faiths.[39] However, Henry L. Stimson stated that at no time did he ever hear Roosevelt, or any other responsible governmental official, suggest that atomic energy should *not* be used to end the war.[40] Winston Churchill was of a similar mind. The decision whether or not to use the atomic bomb to compel the surrender of the Japanese was never really an issue. As Churchill put it, "There was unanimous, automatic, unquestioned agreement around our table."[41]

Sachs's analogy of a "round," however, proved a telling one. The complexity of the Manhattan Project meant that all facets of it had to be readied far in advance of the completion of the various component parts. For example, several buildings at Oak Ridge were finished before the officials knew whether the operations designed to fit inside would actually work. Many decisions had to be made where the unknown factors far outweighed those that were known. Several of the most difficult problems at Los Alamos could not even have been anticipated before the work was well under way.[42] Thus, the leaders of the Manhattan Project had to keep all the balls in the air at the same time.

One of the most crucial of these contingency operations was the famous "Silverplate." This was the army's program to prepare for the combat use of atomic weapons. The Los Alamos dimension of this effort was termed "Project A" (Alberta). The army began their activities in early 1943, and by 1945 the program involved perhaps 1,500 men. Nowhere was the contingency aspect of the Manhattan Project better illustrated.

General H. H. Arnold, Commanding General of the Army Air Force, assumed ultimate responsibility for testing the ballistics of the proposed weapons. The Army Air Force had to know the exact weights and shapes of the atomic bombs. They needed this information, moreover, long before the bombs would be available, at a time when it was uncertain whether they could be made at all, let alone whether they could be made successfully. The technical details, however, had to be frozen by late 1943.

By early 1944, the Army Air Force took a B-29, only recently put into production, and sent it to Dayton, Ohio, where the bomb bay doors were modified to fit the new specifications. Then a special combat team began training for the atomic flights. On August 13, 1944, the unit made its first test drop of an atomic bomb prototype. The initial tests were full of blunders, and the dummy bombs often fell way off target. Months of additional experimentation were needed to ensure reliable fuses, to correct unstable falling patterns, and to perfect their timing. After various design modifications, field tests were resumed in October 1944 and continued on an intermittent basis. Gradually they became monthly, and then almost continuous until August of 1945. All together, a total of 155 test units were dropped between October 1944 and August 1945. In April of 1945, three months before Trinity, the army began preliminary construction on the island of Tinian in the South Pacific. Additional dummy bombs were tested and retested there.[43]

On Saturday, July 14, two days before Trinity, Captain James F. Nolan and Major Robert R. Furman carried the uranium core that was to be dropped on Hiroshima to Albuquerque. From there they flew to San Francisco. The officers then boarded the cruiser *Indianapolis*, and at 8:00 A.M. Monday it set sail for the island of Tinian. It was not until the cruiser arrived at its destination that Nolan and Furman heard of the Trinity results.[44] Such contingency planning was typical of the entire Manhattan Project.

Harry Truman, of course, did not act alone in dealing with the decision to use the weapons in combat. In May of 1945, he sought advice from his secretary of war and, at Stimson's urging, established an advisory Interim Committee of the War Department. Almost alone among the president's inner circle, Stimson possessed a statesman's vision on the significance of the atomic bomb. He realized that the bomb had meaning that extended far beyond the present administration, even beyond the present war. As his diary shows, he reflected an aristocrat's sense of long-range purpose. Accordingly, he urged the president to gather as much expert advice as he could.

After considering the various alternatives—a test drop on an unoccupied island, a "demonstration" with invited Japanese scientists present, a full disclosure of what might happen to Japan if she did not surrender, and so on, the Interim Committee recommended an unannounced military use.[45] Winston Churchill also urged Truman to drop the bombs without prior warning. So, the people whom the president most closely relied on all leaned in the same direction.

It is now known that the numerous second thoughts that were beginning to emerge from the scientific community never reached Truman's desk. The petition from Nobel Prize–winning physicist James Franck, of the Metalurgical Laboratory in Chicago, that from Leo Szilard, those from another group of scientists at Oak Ridge, and others, all remained unread until later.[46] Henry L. Stimson worried a good deal about the bomb, and we know that Truman thought much about it too, although we lack specifics. It has been alleged that Truman called a Roman Catholic bishop in Washington, D.C. to ask how the Catholic Church might respond to the moral consequences of this new weapon.[47]

Truman received advice from numerous quarters, but exactly where and when did he finally make up his mind? The president later told his friend Jonathan Daniels that "Potsdam was where the decision to use the bomb was made."[48] Truman's diary entry for July 25, 1945, reveals that plans had been laid to use the weapon against Japan between that date and August 10. No historian, however, has ever been able to come up with the actual time and place where Truman gave his assent.

Once Truman learned of the enormous success of Trinity, he faced a "decision." What he did was make the simplest decision a person can make. He did nothing. In a sense he made no decision at all. He allowed a process that had been under way in earnest since 1941 to continue unabated.

In a rather inelegant metaphor, General Groves compared Truman to a "little boy on a toboggan."[49] In a sense, the comparison was apt. For example, on the morning of July 23, Stimson cabled George L. Harrison, assistant chairman of the Interim Committee, to find out the time schedule of the bomb and received a response that evening. Truman timed the official release of the Potsdam Declaration—it was dispatched on July 26—so that it would fit into the schedule that Harrison had relayed to Stimson.[50] The schedule, in turn, was based on the production of the fissionable materials at Oak Ridge and Hanford. The tail was

wagging the dog. After the Trinity success, the only action that
Truman might have taken would have been to *stop* the process.
But this was virtually impossible. By August of 1945, the furies
of history, wrapped in the garb of the Manhattan Project, had
assumed a momentum all their own.

Even the official Air Force history of World War II observed
that the command channels that carried out the actual dropping
of the bomb "were highly irregular."[51] The Joint Chiefs of Staff
seem to have been bypassed. Although the ultimate responsibility, of course, lay with the president, he agreed that General
Carl Spaatz, commander of the Strategic Air Forces, and the field
commander directly responsible for delivering the weapon should
be allowed considerable latitude in both the choice of target and
the timing of the mission. Spaatz was instructed to deliver the
first bomb as soon after August 3 as weather allowed. As historian Kenneth Glazier, Jr., has noted, every person, from Truman down to the field commander, delegated part of his authority
to subordinates. Owing to the uncertain weather conditions, the
final choice of the four designated Japanese targets (Hiroshima,
Kokura, Niigata, Nagasaki) ultimately lay with the pilots.[52] This
is hardly the long-range perspective one would hope for in such
endeavors.

The final assembly of the bomb was completed at Tinian on
August 5, but bad weather and the "absence of key crews and
aircraft" postponed immediate action. Clear weather proved vital, for Groves had insisted that the bombs be dropped by actual
sighting rather than by radar. On August 5, General Curtis LeMay
officially confirmed that the mission would take place the next
day.[53] The U-235 bomb dropped on Hiroshima had never been
previously tested, but it reacted just as all the scientists' theories
predicted it would.[54] Both Frank Oppenheimer and Hans Bethe
later recalled that it was not until Hiroshima that they finally
realized the full significance of Trinity. "Hiroshima really brought
it home to me for the first time," Bethe said. "At Alamogordo I
was very much impressed, but I didn't appreciate its meaning."[55]
The Nagasaki plutonium bomb was dropped three days later.

Harry Truman lived until December 16, 1972. Never did he
publicly regret his momentous decision about the dropping of
atomic weapons. He remained scornful of those men, such as
Oppenheimer, who had begun to have second thoughts. Truman

called Oppenheimer a "crybaby" and refused to have anything more to do with him. The president frequently said that he regarded the bomb as simply another military weapon. Momentous decisions, such as Truman was forced into, cannot be reversed, but trying to live up to this image of a "tough decision maker" perhaps pushed him into making statements—"I would do it again"—that might best have been left unsaid.[56] The later Truman became caught up in his own image.

To find Truman's first, unguarded, and probably most honest reaction to the success of the Trinity atomic bomb, one must go to his diary of the Potsdam Conference. This section of his diary was not discovered until 1979; it was published the next year. On July 15, Truman arrived in the Potsdam suburb of Berlin. The conference had been scheduled to start the next day, but Stalin, who was ill, requested a postponement until July 17. Truman used July 16 to tour Berlin. The city that had once been the pride of Germany now lay in utter ruin. As Truman rode through the desolation, he was reminded of other empires— Carthage, Rome, Atlantis, Babylon, and of other rulers, Alexander, Darius, Ghengis Khan—who had fallen in similar fashion. He put the blame squarely on Adolf Hitler—his lack of morals, his folly, his *hubris.* That night he heard of the success at Trinity. With both events in mind, he wrote in his diary: "I hope for some sort of peace—but I fear that machines are ahead of morals by some centuries and when morals catch up, perhaps there'll be no reason for any of it. I hope not. But we are only termites on a planet and maybe when we bore too deeply into the planet there'll [be] a reckoning—who knows?"[57]

Three classic studies illuminate those hectic events that surrounded President Harry S Truman, General Leslie R. Groves, Secretary of War Henry L. Stimson, the members of the Manhattan Project, the Japanese nation, and the Army Air Force pilots from July 16 to August 9, 1945. But they are not what one might expect. The first is Leo Tolstoy's *War and Peace;* the second is Herman Melville's *Moby Dick;* the third is Henry Adams's multivolume *History of the United States during the Administrations of Thomas Jefferson and James Madison.* All three are recognized masterpieces. Each lies equidistant between the heresy that man has no control over his destiny, on one hand, and that man has complete control of it, on the other. These three accounts show man as caught in a set of circumstances over which he has some, but never complete, command. "I plainly confess not to have controlled events," Abraham Lincoln once remarked in 1864, "but that events have controlled me."[58] Harry Truman might well have said, "Amen."

CHAPTER NINE

The Local Legacy

After the July 16 test, many of the scientists collapsed from exhaustion. Both Kenneth Bainbridge and his assistant John Williams found themselves unable to function. They took time off to recuperate through a fishing trip at nearby Elephant Butte Reservoir. Afterwards they returned to the site for several days of gathering data and recovering equipment.

Donning protective gear, Bainbridge led a small crew into the fused sand region to recover the instruments. They were among the first observers to realize the extent to which the ball of fire had depressed the earth. Their Geiger counter readings were so high, however, that each man took only brief turns in digging the equipment out of the sand. The blast had created hot, radioactive gases that often penetrated the shielding devices to ruin measurements. Nevertheless, they did gather valuable information.

The scientists gathered an enormous amount of data. Dozens of experiments provided information on blast pressures, neutron fluences in several energy levels, the emission of gamma-rays, and "generation time"—how the neutron population increased over time—which was closely related to bomb efficiency. Analysis of the distribution of debris led to the formulation of the first fallout model. Virtually every experiment proved significant in one way or other.[1]

The information they collected, however, was of interest chiefly to the scientists. The military concerned itself only with the power of the explosion. Bainbridge coordinated the data and then distilled it in his official report. This was the final duty of most senior officials before they left Los Alamos.[2]

Except for the scientists collecting their data, Trinity Site remained closed to outsiders until eight weeks after the end of the war. The reasons for allowing visitors then were political. Following the Hiroshima and Nagasaki blasts, Japanese spokesmen, and even some American scientists, claimed that the nuclear explosions had made both towns uninhabitable. Columbia University physicist Harold Jacobson declared that any person who visited the two cities within the next seventy years would be committing suicide. The scientists spoke of lingering radioactivity; the Japanese civilians talked simply of "poison."[3]

The army denied this allegation. It maintained that no radioactivity remained in the two cities. Army spokesmen insisted that the only radiation released came at the instant of detonation and that this soon disappeared.[4] In fact, General Thomas Farrell specifically instructed the first Atomic Bomb Investigation Committee to Japan that their mission was to prove that there was no radioactivity left from the bomb.[5] (They insisted on measuring accurately.)

General Groves reasoned that if American newsmen and photographers could visit Trinity Site for themselves, they would accept the army's position. So, on the afternoon of Sunday, September 9, 1945, thirty-one writers and photographers, escorted by Groves and several other officers, plus five scientists, spent about half an hour at Ground Zero. This widely publicized visit brought Trinity onto the front pages of the newspapers and magazines. *Life* gave the visit thorough coverage.[6] For the first time, Trinity Site received the publicity that had been denied it by the war.

The group lunched on baked chicken with the fifty men stationed at the MP Base Camp and heard atomic bomb stories from newly promoted Captain Howard C. Bush. Afterwards, they donned protective foot covering and climbed into jeeps to head for Ground Zero. As they approached the region, they saw the fire-blackened skeletons of greasewood plants, all leaning away from the direction of the blast. They then arrived at an area, almost 4,800 feet across, where all the grass and greasewood had disappeared. Oppenheimer remarked that it looked as if it had been dragged by a scraper. Jumbo resting on its base was pointed out as representing the scientists in a "pessimistic mood."[7]

The group arrived at the crater site almost before they realized it, for it was hardly noticeable. It differed from every other crater on the face of the earth. Instead of gouging out the ground and blowing it away, as most newsmen had expected, the pressure from the blast had simply pushed the earth straight down. This produced only a gentle, saucer-like slope whose center lay ten feet deeper than the edges, about 1,200 feet away. As a Park Service employee described it two months later, "This [depression] is so very slight, compared to the acreage involved, that, if the area were unfenced and the ground were not covered with 'atomsite,' one might ride or walk directly through it without even suspecting that any unusual happening had ever taken place there."[8]

Everyone marveled at the layer of trinitite, which resembled "a sea of green."[9] It stretched out to an average radius of 400 yards, with several spurs extending 100 yards farther. Many of the newsmen picked up pieces as souvenirs. The fused sand came in fantastic shapes and sizes. Mixed with it lay some pieces of reddish glass, brought on by the presence of iron or copper in the vicinity, and these were especially prized. The reporters also found numerous small lumps of earth that had been completely encased in a trinitite coating, the thickness of an eggshell. These pieces were estimated to have fallen from the mushroom cloud, for they proved to be very radioactive. Each souvenir was carefully checked by the Geiger counter, and some were discarded as being too hot to take home.[10]

The use of radiation counters also fascinated the newsmen. They explained to their readers that the items that "looked like camera boxes" were really devices to check radioactivity. Louis Hempelmann explained that the activity in the area was around 12 R per hour, but that the effect was believed to be cumulative. A total of 600 R was considered lethal. Hempelmann explained the idea of half-life and assured them that the activity was decreasing at the rate of 50 percent a month. Kenneth Bainbridge discussed how some people had recommended that the army try to gather up all the trinitite from the region and extract the plutonium from it. The idea has been dismissed as impractical.

The visit was kept brief. Everyone wore protective white boo-

ties to keep out radioactive dust and the cameramen were told
that if they remained over forty-five minutes, they might find
their film fogged. When Hempelmann walked up to the base of
the tower, where Oppenheimer had been standing for too long
a period, and whispered "eighteen," he had considerable explaining to do for the newsmen.[11]

The trip produced much publicity and several now-classic photographs. As far as the army was concerned, it served its purpose
admirably. Stories of lingering radioactivity on Japanese soil disappeared from the papers. Stafford Warren's survey team, sent to
investigate the radiation at Hiroshima and Nagasaki, arrived at
similar conclusions. In fact, Trinity Site proved hotter than either
Japanese city. The bombs there were exploded 2,000 feet in the
air while the Trinity fireball actually touched the ground.[12] As
there was no loss of life at Trinity, other comparisons have little
meaning.

Within a short period, the citizens of New Mexico suddenly
became very proud of Trinity. On August 11, 1945, representatives of the Alamogordo Chamber of Commerce urged that the
site be made into a national monument. "People will want to
see the spot where this great event occurred bringing in a new
era," they wrote, "the same as they have wanted to see the spot
where the first Pilgrim set foot on the continent."[13] Governor
John J. Dempsey, Senator Dennis Chavez, and numerous others
joined in. Charles S. McCollum of Las Cruces wrote Senator
Carl A. Hatch and urged him to begin the process. McCollum
hoped that gigantic buildings would be erected at Trinity and
urged that the surrounding area be made into a peace park.[14]

Secretary of the Interior Harold Ickes instructed the National
Park Service in Santa Fe to survey the region immediately. Security restrictions were such, however, that they could not even
gain entry until early November 1945. After the visit, the Park
Service drew up an extensive recommendation that an area of
2,865 acres be set aside as a national monument.[15] Their blue-
prints included a large monument headquarters, parking for two
hundred vehicles, concession stands, a sheltered walk through
the trinitite to Ground Zero, and numerous appropriate photographs and signs.[16]

Secretary Ickes also prepared an elaborate proclamation on the
subject. It stated that the atomic bomb had reduced the loss of

life in World War II; that it represented the united efforts of American and British science, and American industry and labor; and that it harnessed the "basic power of the universe." Ickes hoped that atomic power could be used both to maintain world peace and to create a higher standard of living for all the world. The proposed national monument would commemorate "that great historic and scientific event."[17] The Park Service even drafted a similar statement for Harry Truman to use when he officially gave the area monument status. Truman, however, never did so, and his unspoken declaration now rests in the Park Service archives in Santa Fe.

Local editors praised the proposed monument idea. They hoped that the state would benefit from the increased tourism. They noted that visitors to White Sands National Monument could also stop at Trinity with little difficulty.[18] New Mexico's largest newspaper, the Albuquerque *Journal*, enlisted as one of the strongest supporters of the project. A Trinity National Monument, it asserted, would long keep the state in the public eye.[19]

The editors of the *Journal*, however, also noted that there might be some problems with the monument. Trinity Site lay about fifty miles from any town, and the roads into it were abysmal. The layer of trinitite, once so impressive, was already beginning to erode, and the shallow depression at Ground Zero did not provide an especially memorable image. The editors recommended that a detailed scale model of Trinity Site be built, and that this, along with numerous appropriate photographs, be placed in a more accessible location.[20] They doubtless meant Albuquerque.

The State Land Commissioner, John Miles, and several other state officials viewed the proposed monument from a different perspective. They feared that the monument might lock up thousands of acres of land. The Jornada, they said, was not true desert but proved quite usable for grazing. The state wanted to get it back in production as soon as possible. Senator Hatch had to assure state officials that the proposed monument would cover only a small area, and that the rest of the acreage would soon be back in use. Everyone assumed that the Alamogordo Bombing Range would be dismantled.

Such hopes were not fulfilled. The army maintained that it needed to keep this property for defense reasons. Thus, it demonstrated no eagerness to return it. The National Park Service survey team of 1945 acknowledged the reality of the situation. They recommended that interim measures be taken to preserve all the historic features and buildings of the area until it could be officially released from security restrictions. At this time, the Park Service foresaw that it would take over and restore Trinity as a monument.

The following year, the Park Service pushed again for a monument, but the War Department negated the proposal, again on the basis of military necessity. The Park Service tried again in 1948 and 1950–52, but the results were the same. By the early 1950s, the old Alamogordo Bombing Range had evolved into the 4,000-square-mile White Sands Missile Range. Ironically, as security relating to Trinity Site became less and less, the security around the Missile Range increased dramatically.[21]

In the meantime, responsibility for the site had changed hands. On January 1, 1947, much of the land around Ground Zero was transferred to the Atomic Energy Commission, which remained responsible for the radiation safety of the region. The AEC said little about the proposed monument and did not encourage visitors. They noted that continued studies of the region were under way to test "the cumulative effects of radiation." No one was willing to say how long such studies would continue. The National Park Service told the newspapers that it was confident that the area would become a monument some day and stated that for the interim they were satisfied to leave the protection of the area to the army. "There is no rush," one official said.[22]

Privately, however, the Park Service was very unhappy with how the army was preserving the various structures at Trinity: Jumbo, the numerous bunkers, Base Camp (now termed *Camp Trinity*) and especially the McDonald ranch house. This old stone building, with its numerous outbuildings, had long fallen into decay. Moreover, the Park Service accused the army of bulldozing some bunkers simply to avoid repairing them. It also charged that lax security had allowed vandals to penetrate the area and shoot holes in the various structures.

In 1952, the AEC, working with the army, had actually planned to cover the entire bomb crater with fresh dirt or asphalt. The trinitite had weathered to such an extent that it was becoming a menace. The wind easily picked up the fine dust and formed drifts, which began to concentrate the radioactivity. Only sharp protests from New Mexico's congressmen halted this plan.[23] As a compromise, the trinitite was packed into barrels and buried elsewhere. A small portion of it was preserved and protected under a roofed shed.

In March 1952, New Mexico Congressman Antonio Manuel Fernandez (D) introduced a House bill to establish a Trinity Atomic National Monument. This created several months of paperwork among various governmental agencies as to the safety and general condition of the area. Thomas L. Shipman, health division leader at Los Alamos, assured everyone that Trinity Site presented no danger to any potential visitors. Still, nothing was done.[24]

Finally, in 1965 the Park Service had Trinity Site declared a National Historic Landmark. In 1975 they had it declared a National Historic Site. One provision of such designations was a yearly inspection by a Park Service official, and the 1973 Park Service report by Theodore R. Thompson accused the army of letting things go to ruin, either through neglect or willfully.[25]

This led to some rather testy exchanges until the Park Service and the White Sands Missile Range reached an understanding. The army declared that it considered Trinity Site to include only the fenced-in crater around Ground Zero (where nothing existed), and they stoutly maintained that they had always kept it in its original condition. The Park Service argued that Trinity Site meant all the bunkers, outbuildings, poles, Jumbo, and the McDonald ranch house. Finally in 1980, the army agreed to stabilize those bunkers that had fallen into disrepair. Because of restrictions on the original McDonald lease, however, they declared that no improvements could be made there until they had assumed full title.[26]

With this new status and publicity, both Los Alamos and the army began to take measures to ensure the safety of any potential visitors to the site. In March 1967, the army bulldozed two bunkers that had become dangerous. The same year Los Alamos sent a crew down to detonate an older, moderately radioactive charge (nicknamed "Sleeping Beauty") that had failed to explode in 1946 because of a faulty switch. The men bulldozed open the bunker, repaired the switch, detonated the explosives, and sealed it up again.[27] In addition, they dug up numerous garbage cans full of

buried radioactive material. These were placed in fifty-five-gal-
lon drums and shipped to Los Alamos for further study. Over
1,200 readings were taken on the trinitite, and Los Alamos concluded that the glossy pieces (hence the most desirable as souvenirs) had lower readings than the duller ones. But they decreed
all the readings acceptable.

Finally, Frederic L. Fey, Jr., of Los Alamos came down to do a
thorough health-physics survey of Trinity. All the various outbuildings had previously been found free of radioactivity, so Fey
concerned himself solely with the fenced-in area. While he warned
that the trinitite should not be removed by visitors, he considered the area "basically" safe. A person would have to eat 10^4
grams of trinitite or hold a piece of it against his skin for eighty-
three straight days to receive any type of radiation injury. Fey
concluded that "it does not appear that anyone could receive
any radiation injury through a visit there." Still, he urged that
all Park Service personnel wear radiation badges and that long-
term exposure records be kept for them.[28]

With the increased environmental concern of the 1970s, scientists made two more health surveys of the Trinity area. In
September 1972, a team from Los Alamos took numerous soil
and vegetation samples throughout the region. It had been twenty-
seven years since the explosion and seventeen since the last
survey of the area. The main change that the Los Alamos team
discovered was that the plutonium had continued to migrate
deeper into the soil. They found that plutonium, which had been
deposited on the surface of the earth, had penetrated at least 30
cm underground in several spots.

In 1973 and 1974, the Environmental Protection Agency (EPA)
surveyed the plutonium levels throughout a much wider (4,500
square miles) area of central New Mexico. This team took eighty-
eight soil samples—many on the Chupadera Mesa—plus several
months of air samples at the Monte Prieto ranch near Socorro.
While the plutonium was easily measurable in the surface 5 cm
of soil, they concluded that the amount discovered was less than
half of the Federal Radiation Guidance "screening level" per square
meter of earth. The airborne contamination was a factor of 25
below the EPA guidelines. Thus, there seemed little cause for
worry, should there be any change in the status of Trinity Site.[29]

Since 1945, the National Park Service has frequently waxed

enthusiastic about the proposed Trinity National Monument. By general consensus, Trinity Site stood out as the most significant of all those locations connected with the Manhattan Project. The Park Service denied that the atomic displays at Alamogordo, Oak Ridge, Los Alamos, and Albuquerque obviated the need for a monument at Trinity.

The Park Service has devoted many hours to the idea of a Trinity National Monument. Their files contain numerous proposals for it. These include: bringing in a B-29, an inert model of "fat man," restoring several *T* poles and wires, seeding Ground Zero with a facsimile of trinitite, laying numerous self-guiding trails, restoring all bunkers with self-operated interpretative devices, and the stationing of numerous rangers.[30] Nearby Gran Quivira National Monument received 12,000 visitors in 1970; the Park Service predicted a Trinity National Monument might reach 150,000 visitors a year.[31]

Others have been noticeably less enthusiastic. While few people would deny that Trinity Site is *significant* enough to be included within the National Park system, there are those who question how *appropriate* it might be. In 1969, for example, a Denver *Post* reporter suggested that Trinity was America's guilt symbol, similar to a German concentration camp.[32] Norris Bradbury argued that giving Trinity national monument status would create endless controversy.[33] Every July 16 would be a potential day of disturbance.

Officials from the White Sands Missile Range have been equally cool to the monument idea, but for different reasons. The proposed Trinity Monument covers 51,500 acres—Ground Zero, all the N-, S-, and W-10,000 bunkers, the numerous other buildings, and the McDonald house. It lies on land that is currently used for the testing of missiles. One Park Service survey of Trinity frankly admitted that a stray missile might sometime hit a historic building. The thought of 150,000 tourists roaming across forty miles of desert road, 365 days a year, would drive a White Sands Missile Range official to despair.

In 1967, New Mexico's senators introduced a bill authorizing a transfer of Trinity Site to the Park Service and its opening to the public. This would happen whenever the missile range at White Sands was no longer needed by the Department of Defense. When this might occur, however, is another question. As

a Park Service official sadly noted in 1979, "There is no indication that White Sands will be anything but essential for military uses in the future."[34] The army agreed to watch over the
various structures, to make certain they do not deteriorate any
further. With this as a compromise, the Park Service has shelved,
it appears permanently, all attempts to turn Trinity Site into a
national monument.

Another item of considerable local concern involved Jumbo.
The huge container had been placed within a seventy-foot-tall
steel tower, about half a mile from Zero. The blast levelled the
tower—an indication of what might happen to a steel-beam
building—but Jumbo emerged unscathed. It remained there for
almost two years. Then, in April of 1947, General Groves began
to worry that a congressional investigating committee might
view an untested Jumbo as a gigantic waste of taxpayers' money.
Accordingly, Groves instructed several junior officers at the
Special Weapons Division of Sandia Base in Albuquerque to explode it. Without informing their colleagues, these men took
several 500-pound bombs and placed them inside the vessel,
resting them on the bottom hemispheres. Since the explosive
force was not placed in the correct position (the exact center),
the blast blew out both of Jumbo's ends.

When Robert W. Henderson, Roy W. Carlson, and the officials
at Babcock and Wilcox heard what had happened, they were
furious. Placing conventional bombs in the wrong place hardly
provided a valid test of the vessel's design capabilities. As Henderson recalled later, "It was an absolute scandal."[35]

After this debacle, the army unceremoniously bulldozed Jumbo
into a nearby trench. It lay underground until 1951, when a team
of scientists disinterred it for possible use in another series of
explosive tests. After examining its remains, however—now down
to about 180 tons—they concluded that it would not meet their
needs. So, Jumbo was abandoned again, this time for nine years.

In the spring of 1960, the Socorro Chamber of Commerce became interested. They began inquiry to see if they could move
Jumbo to the city and make it "the featured attraction" of their
proposed Atomic Memorial Park.[36]

Socorro already had a small monument to the world's first
nuclear blast in their downtown square (where, ironically, an

end piece from Jumbo was exhibited, mistakenly identified as a piece of the original atomic bomb). But the plans for this park were extensive. It was to boast lighted ballfields, tennis and volleyball courts, and elaborate picnic areas. Since access to Ground Zero was restricted, Socorro officials hoped this easily accessible park would capture the public's growing fascination with atomic energy. "The park," declared the Socorro *El Defensor Chieftain*, "would be a mecca for tourists as well as a big boon for local residents."[37]

The army seemed delighted to have Jumbo off its hands. After a year of paperwork on the question of actual ownership of the vessel, they officially donated it to the city of Socorro. There was only one condition: Socorro officials would have to move it from White Sands Missile Range property.[38]

Socorro townspeople were delighted, and Preston K. Pond, a Missile Range employee, as well as head of the Chamber of Commerce, began a campaign to whip up the necessary enthusiasm. Pond posed for several publicity photos next to (and inside of) his main attraction, and people across the nation responded. After years of neglect, Jumbo suddenly captured the public's eye.

Numerous free-lance writers and photographers came to Socorro to tell the tale. While the *Los Alamos Scientific Laboratory News* described Jumbo as "a monumental bad guess," that became decidedly a minority position. A San Diego reporter spoke of Jumbo as "lonely, neglected, homeless."[39] Senator Clinton P. Anderson (D-NM) told his colleagues that the city of Socorro was only trying to give an old World War II "veteran" a home. Reporter Howard Simons of the Washington *Post* pulled out all the stops. He wrote of the "big fellow" who had hoped "to swallow America's first atomic bomb" but who now "wandered in the desert—forlorn and forgotten."[40] Jumbo, Preston Pond, announced, still had "hundreds of years of life" left in it.[41]

In spite of this enthusiasm, however, the Chamber of Commerce's plans came to naught. City officials made numerous inquiries to both public and private haulers, but no one could discover a way to carry the 180-ton container across the lone Rio Grande bridge that separated Trinity Site from Socorro. So, Jumbo had to remain where it was. It was not until the early 1970s that Missile Range officials rolled it to just outside the fenced-in area at Ground Zero and erected a small plaque. There it still lies today.[42]

The significance of the original Trinity Site has not been widely appreciated. Only recently has the White Sands Missile Range acknowledged its historic importance. For almost twenty years following the detonation, the blast area was identified only by a worn wooden sign marked "Zero." It was not until 1965 that the army erected a modest marker at the site of Ground Zero. In 1953, however, Missile Range officials took a major step. Working closely with the Alamogordo Chamber of Commerce, they decided to open the site to visitors one day each year. Because of the often unbearable weather in July, the army selected the first Saturday in October.

About 600 people arrived the first year the area was opened, and the number has grown steadily. Today about 3,000 people make the trip annually. All visitors are escorted into the region via automobile caravan, and the trip is restricted to a total of ninety minutes. Visitors travel only to Ground Zero. They are not allowed to visit any of the outlying bunkers or the dilapidated McDonald ranch house. An army brochure prohibits people from eating or drinking within the fenced-in area, for fear of ingesting radioactive dust particles. Pregnant women and small children are discouraged from making the trip.[43] The arrangements are efficiently handled.

Trinity Site is not as impressive as one might expect. The landscape is bleak, and only the small monument distinguishes the fenced-in crater from the surrounding countryside. The depression of the ground caused by the blast is hardly noticeable. Indeed, if one had not been told beforehand that something memorable had happened here, nothing in the area would give it away.

Almost all the fused sand has been carted away and buried. A long, low wooden shed has been built over the last remaining section of the trinitite in a halfhearted attempt to preserve it in its original condition. But the shed is slowly filling up with the drifting sand that every windstorm carries with it. One gains little impression of how the area must have looked immediately after the blast. The desert is slowly reclaiming its own.

It is somewhat ironic that the most memorable permanent object at Trinity Site is Jumbo, *sans* ends. Located just outside the fenced-in area, next to a small marker, the huge container is the first object one encounters on the path to Ground Zero. Small children run up and down its walls, and every year hundreds of people take its picture. Although Jumbo was never officially utilized at the Trinity test, by a strange quirk of fate it has become the most widely recognized item connected with the event.

Numerous officials from the Los Alamos National Laboratory, the White Sands Missile Range, the National Park Service, and the National Atomic Museum in Albuquerque provide information to the visitors. Their photographic exhibits, booklets, and speeches attempt to recapture the significance of the site. In 1982, museum representatives displayed a life-sized model of the Nagasaki plutonium bomb. Perhaps the most memorable comments that year, however, came during the introductory prayer for peace delivered by the range chaplain. The burden of his message was: "Lord, we have abused the gifts which Thou hath given us."

By opening Trinity Site to visitors once a year, the army has reached a realistic compromise on the national monument issue. Although the Park Service remains ever hopeful, it is unlikely that a national monument will be established at Trinity within the foreseeable future.

Yet perhaps Trinity Site left in its natural state speaks more eloquently than any monument could. The ambiguity of the events staged there has made unanimous celebration of them difficult.[44] It is better perhaps that they remain chiefly in the mind's eye. There each person may reflect, without contradiction, on that morning of July 16, 1945, when the sun rose twice in the New Mexico sky. And when it did, it changed the world forever.

Epilogue

After the war, the majority of the Los Alamos scientists—"the first team"—left the Hill to resume their academic careers. The town itself was filled with numerous petitions for world government, unilateral disarmament, and the internationalization of the atom. Hopes ran high that the fledgling United Nations could somehow ensure international cooperation on nuclear matters.[1]

At the same time, the scientists found themselves public figures. Oppenheimer's picture graced numerous popular magazines, and he became virtually the symbol of the new status of science in American life. Many others, such as Hans Bethe, I. I. Rabi, Victor Weisskopf, and Robert Wilson, devoted countless hours detailing the significance of their discoveries to the nation.[2] "No dinner party is a success," observed *Harper's*, "without at least one physicist to explain . . . the nature of the new age in which we live."[3]

The scientists took their new responsibility seriously. In the immediate postwar years, they were the only ones who really understood the full implications of atomic weapons. For the rest of the nation, comprehension came slowly. The gap may be seen in the following incident. A few days after the Japanese surrender, General Carl Spaatz flew to the island of Tinian and asked to see the working parts of an atomic bomb. When physicist Charles Baker showed him a mock-up of the nuclear device, Spaatz was astounded. "Young man," he said, "do you mean to tell me that what went in that place that you're showing me there made all that explosion?" Baker replied that it did. "Well, young man," Spaatz continued, "you may believe that but I don't."[4]

The scientists utilized their new influence with skill. It was largely due to their speeches and lobbying efforts that America's nuclear affairs were placed under the civilian-dominated Atomic Energy Commission, rather than given exclusively to the military, as had been originally recommended. They also helped formulate what eventually became the Baruch plan—a complex proposal for the international control of nuclear materials; the Baruch plan would have established almost an "international Manhattan Project." The Soviet Union would have no part of this, however, and it soon collapsed. In 1949, the Soviets exploded their first atomic bomb. They detonated a thermonuclear bomb in 1953. The arms race had begun in earnest.

As might be expected, the Los Alamos scientists fanned across the entire spectrum on such political issues as the internationalization of atomic power and the need for further weapons development. Perhaps the most visible reaction, however, came from Oppenheimer himself. In a November 16, 1945, speech in Los Alamos, where he accepted the Army-Navy "Excellence" award for the town, Oppenheimer became almost prophetic. If the atomic bombs were to be added as new weapons to the armaments of a warring world, he said, then the time will come when mankind will curse the name of Los Alamos.[5] In 1948, in a widely reprinted article, he confessed that "the physicists have known sin; and this is a knowledge which they cannot lose."[6] Even after the Ulam-Teller discovery made the proposed hydrogen bomb "technically sweet," he remained dubious that it should be developed.

Yet not all the Los Alamos scientists felt the same anguish and guilt. As Freeman Dyson has noted, many of the Cornell physics department resented Oppenheimer's confession. They wanted no one to weep in public for their actions. They considered themselves no more guilty than anyone else who had made lethal weapons during World War II.[7]

Those who remained at Los Alamos, under its new director, Norris Bradbury, felt the same way. After the failure of all international plans of control, combined with Soviet advances in weapons, this position gained in popularity. Several original

Manhattan Project people remained to work on the Hill, and they were often joined by other "first term" members in the capacity of "consultants." As Harold M. Agnew, later to be Bradbury's successor as lab director, remarked in 1954: "There is not now and there never has been, any feeling of guilt toward what we are developing or toward the goals to which we are working."[8]

The most vocal conservative spokesman has been Edward Teller. During the 1950s, Teller became widely known for his insistence on the need for both a hydrogen bomb and a strong defense program. In 1952, the Lawrence-Livermore Laboratory in California was established to further his position. The military supported Teller and argued that peace could only be assured through the maintenance of strength. They liked to quote Dean William Inge, of St. Paul's Cathedral in London, who noted during the late 1930s, when Germany began rearming, that "it is useless for the sheep to pass resolutions in favor of vegetarianism as long as the wolf remains of a different opinion."[9]

The liberal wing, however, developed a much wider following. Several of the scientists returned from the war as pacifists. Others, such as Robert Wilson and Victor Weisskopf, vowed never to work on nuclear weapons again. Their most prominent spokesman became George Kistiakowsky. Appointed science advisor to Dwight Eisenhower in 1959—just as the two superpowers had begun to increase their missile delivery systems—Kistiakowsky steadfastly endorsed a program to limit armaments.[10] He remained active in such endeavors until his death on December 7, 1982. The next year Nobel laureate Owen Chamberlain noted that the world's nuclear arsenal could be reduced 95 percent without either side falling below a sufficient level of deterrence.[11]

By the early 1980s, the world had witnessed over one thousand atomic explosions. Estimates place the current global arsenal at perhaps 100,000 nuclear warheads. The "nuclear club" includes the United States, Great Britain, France, the Soviet Union, India, China, South Africa, and (almost surely) Israel. The prospects that Brazil, Argentina, or Pakistan might soon join this elite circle have been often discussed. The initial Trinity gadget weighed several thousand pounds and proved awkward to handle. Modern weapons dwarf that first explosion by a factor of thousands. One hydrogen bomb could destroy New York City. A bomb equal to that detonated at Trinity can now be carried in a suitcase.

Millions of people around the globe have reacted to this awkward situation. Massive public demonstrations in Germany, Italy, England, and the United States, however, have had minimal impact on changing the policies of the political leaders. Solitary protests by such Russian scientists as Andrei Sakharov, the "father" of the Soviet H-bomb, have had even less effect.[12] Ironically, at a June 1983 antinuclear demonstration in Albuquerque, one of the people involved was Peter Oppenheimer, Robert's son. Perhaps a culmination of this feeling of frustration occurred in the previous spring when a philosopher at a well-known Southwestern university publicly stated that he wished the atom had never been split.

At the fortieth anniversary of Los Alamos—April 1983—eighty-four-year-old I. I. Rabi tried to distinguish between the atmosphere of the 1980s and that of the 1940s. In a major address, he said that the misuse of the Los Alamos scientists' creation during the last forty years should not burden the creators of the original weapon with feelings of guilt.

"What we did was great," he told an appreciative audience; "what we did was inevitable; what we did was fortunate for the United States, and for the world—as of that period." Rabi and his colleagues remain convinced that if the Nazis had developed the atomic bomb first, the whole of Western civilization would have been drastically altered.[13]

But over the years, Rabi continued, the scientists have placed their discoveries in the hands of political leaders. They, in turn, did not fully understand the power of what they possessed. When Sakharov once approached the Soviet head of state on this matter, he was rebuffed with a Russian parable to the effect that the Czar did not need the advice of the peasants. Since 1945, Rabi suggested, the scientists had abdicated the responsibility that came with their superior knowledge.

Some scientists feel that the generation of the 1980s has less respect for nuclear weapons than the generation of the 1940s. The World War II scientists literally built the bombs with their own hands. They were familiar with all the features of their creation, from the ground up. Few people today can boast this familiarity.[14] Mathematician J. Carson Mark has half seriously suggested that all heads of state in the United Nations be forced to witness an atmospheric nuclear explosion once a year.[15] This lack of appreciation of the power involved, Rabi concluded, has created an atmosphere where human beings are considered "as if they were matter" with "nations lined up like those prisoners at Auschwitz."[16]

What happened at Trinity Site, therefore, has evolved into the most crucial issue of the twentieth century. The creation of the Los Alamos scientists, Victor Weisskopf observed at the Los Alamos anniversary, "has become the unintended cause of the world's tragic predicament." "Forty years ago we meant so well," Weisskopf continued, "but it did not turn out so well."[17]

Abbreviations

COHP	Columbia Oral History Program, Columbia University, New York, New York
FDR	Franklin D. Roosevelt Library, Hyde Park, New York
HST	Harry S Truman Library, Independence, Missouri
LAHS	Los Alamos Historical Society
LANL	Los Alamos National Laboratory Archives
LASL	Los Alamos Scientific Laboratory (From c. 1945, the Laboratory was known as LASL; after 1980 it became LANL.)
MED	Manhattan Engineer District
NAM	National Atomic Museum, Albuquerque
NMSRA	New Mexico State Records and Archives, Santa Fe
NPS	National Park Service, Southwest Division, Santa Fe
SHS	Socorro Historical Society
UCLA OHP	Oral History Program, University of California, Los Angeles

Notes

Introduction

1. Mrs. Bernice Brode, "Life at Los Alamos, 1943–45," *Atomic Scientists' Journal* 3 (November 1953): 89.
2. Interview with Louis Hempelmann, M.D., December 22, 1981, Santa Fe, New Mexico.
3. William Laurence, *Dawn Over Zero: The Story of the Atomic Bomb* (New York: Alfred A. Knopf, 1946); Lansing Lamont, *Day of Trinity* (New York: Atheneum, 1965); Kenneth T. Bainbridge, *Trinity* (Los Alamos: Los Alamos Scientific Laboratory, 1976). Some parts of the Bainbridge final report still remain classified.

Chapter One

1. The special issue of *Scientific American* 183 (September 1950) details these advances in numerous fields.
2. Robert L. Weber, *Pioneers of Science: Nobel Prize Winners in Physics* (Bristol and London: Institute of Physics, 1980), quoted 188.
3. Robert Jungk, *Brighter Than a Thousand Suns: A Personal History of the Atomic Scientists* (New York: Harcourt Brace Jovanovich, 1958), 3–70; Ronald W. Clark, *The Scientific Breakthrough* (New York: G. P. Putnam's Sons, 1974), 146–68; G. E. M. Jauncey, "The Early Years of Radioactivity," *American Journal of Physics* 14 (July–August 1946): 226–41; Noel Pharr Davis, *Lawrence and Oppenheimer* (New York: Simon and Schuster, 1968), 11. Daniel J. Kevles, *The Physicists* (New York: Alfred A. Knopf, 1978) is the best story of the profession as a whole.

4.   The best example of this can be seen in the "Letters to the Editor Section" of the major physics journal, *Physical Review* 58 (July 1, 1940). Both Flerov Petrjar of the Physico Technical Institute (F), Radium Institute (P) of Leningrad, and Kwai Umeda of the Department of Physics of Hokkaido Imperial University, Sapparo, Japan wrote in to share their discoveries, 89–91.

5.   Laura Fermi, *Illustrious Immigrants: The Intellectual Migration from Europe, 1930–1941* (Chicago: University of Chicago Press, 1968), 11. See also Donald Fleming and Bernard Bailyn, eds., *The Intellectual Migration: Europe and America, 1930–1960* (Cambridge, Mass.: Harvard University Press, 1969). Gerald D. Nash has a volume in preparation, *The West Transformed: The Impact of World War II*, which will deal with the impact of these immigrants on the American West.

6.   Charles Werner, "A New Site for the Seminar: The Refugees and American Physics in the Thirties," in Fleming and Bailyn, *The Intellectual Migration*, 190–228.

7.   Otto R. Frisch, "How It All Began," *Physics Today* 20 (November 1967): 43–52; Otto Hahn, "The Discovery of Fission," *Scientific American* 198 (February 1958): 76–84.

8.   William L. Laurence, "The Atom Gives Up," *Saturday Evening Post* (September 7, 1940): 60.

9.   James P. Baxter, *Scientists Against Time* (Boston: Little, Brown and Company, 1946), 420.

10.   William L. Laurence, "Vast Power Source in Atomic Energy Opened by Science," *New York Times* (May 5, 1970) 1: 1; I. I. Rabi, *Science: The Center of Culture* (Cleveland: New American Library, 1970), 13.

11.   S. M. Ulam, *Adventures of a Mathematician* (New York: Charles Scribner's Sons, 1976), 5.

12.   "Before the War we physicists almost never had occupied ourselves with problems and questions which could in any direct way be called immediately practical," confessed I. I. Rabi. *Science: The Center of Culture*, 1.

13.   Albert Einstein to Franklin D. Roosevelt, August 2, 1939, Atomic Bomb Folder, Franklin D. Roosevelt Library, Hyde Park, New York; Ronald W. Clark, *Einstein: The Life and Times* (New York: Avon Books, 1971), 659–81; Leo Szilard, "Reminiscences," in Fleming and Bailyn, *The Intellectual Migration*, 112–73.

14.   Nat S. Finney, "How F.D.R. Planned To Use the A-Bomb," *Look* 14 (March 14, 1950): 24–26.

15.   Stephane Grioueff, *Manhattan Project: The Untold Story of the Making of the Atomic Bomb* (Boston: Little, Brown and Company,

1967), 10; James MacGregor Burns, *Roosevelt: The Soldier of Freedom* (New York: Harcourt Brace Jovanovich, 1970), 249–52. See also, Lee Bowen, *A History of the Air Force Atomic Energy Program, 1943–1953. Introduction and Chapter I (1959): Project Silverplate, 1943–1946*, typescript, National Atomic Museum, I, 58.

16. The best books on the Manhattan Project are: Richard G. Hewlett and Oscar E. Anderson, Jr., *The New World 1939–1947: A History of the United States Atomic Energy Commission* (University Park, Pa.: Penn State University Press, 1962); and Stephane Groueff, *Manhattan Project*. The official report, released immediately after the war, is also good: Henry DeWolf Smyth, *Atomic Energy for Military Purposes: The Official Report on the Development of the Atomic Bomb under the Auspices of the United States Government, 1940–1945* (Princeton: Princeton University Press, 1947). Anthony Cave Brown and Charles B. MacDonald have put together an edited version of the government's massive official history of the project in *The Secret History of the Atomic Bomb* (New York: Dial Press, 1977).

17. Spencer R. Weart, "Scientists with a Secret," *Physics Today* 29 (February 1973): 23–30.

18. Carroll L. Wilson to V. Bush, Feb. 24, 1943, in Atomic Bomb Folder, FDR; Vannevar Bush, *Pieces of the Action* (New York: William Morrow and Company, 1970); Lawrence Badash, "Introduction," in Lawrence Badash, Joseph O. Hirschfelder, and Herbert P. Broida, eds., *Reminiscences of Los Alamos, 1943–1945* (Dordrecht, Holland: D. Reidel Publishing Company, 1980), xiv. A. Hunter Dupree, "The Great Instauration of 1940: The Organization of Scientific Research for War," in Gerald Holton, ed., *The Twentieth-Century Sciences: Studies in the Biography of Ideas* (New York: W. W. Norton, 1970), 456.

19. Leslie R. Groves, *Now It Can Be Told: The Story of the Manhattan Project* (London: André Deutsch, 1963).

20. Enrico Fermi, "The Development of the First Chain Reaction Pile," *Proceedings of the American Philosophical Society* 90 (January 1946): 20–24; *Albuquerque Journal* (November 28, 1982). Herbert L. Anderson, "The Legacy of Fermi and Szilard," *Bulletin of the Atomic Scientists* 30 (September 1974): 56–62, and (October 1974): 40–47.

21. Eugene P. Wigner, *Symmetries and Reflections: Scientific Essays of Eugene P. Wigner* (Bloomington: Indiana University Press, 1967), 238–44; Leona Marshall Libby, *Uranium People* (New York: Charles Scribner's Sons, 1979), 121. There is now a plaque on the entrance to the squash court that reads: "On Dec. 2, 1942 Man achieved here the first self-sustaining chain reaction and thereby initiated the controlled release of nuclear energy."

22. Fermi's recollections, originally published in the Chicago *Sun-*

Times (November 23, 1952), have been reprinted in Corbin Allardice and Edward R. Trapnell, *The First Reactor* (AEC pamphlet, c. 1957). See also Enrico Fermi, "Elementary Theory of the Chain-reacting Pile," *Science* 105 (January 1947): 27–32.

23.   James P. Baxter, *Scientists Against Time*, 432.

24.   Interview with Marvin Wilkening, January 12, 1982, Socorro, New Mexico. "Pre-historic beings like me," Laura Fermi wrote in 1970, "consider the first date [December 2, 1942] the only true birthday of the atomic age." Laura Fermi, "Bombs or Reactors," *Bulletin of the Atomic Scientists* 26 (June 1970): 29.

25.   Charles W. Johnson and Charles O. Jackson, *City Behind a Fence: Oak Ridge, Tennessee, 1942–1946* (Knoxville: University of Tennessee Press, 1981); there is no comparable story of Hanford.

26.   Laurence, "The Atom Gives Up," 62.

27.   Leslie R. Groves, "The Story of the Atomic Bomb," *Think* [IBM Publication] 11 (November 1945): 5–7, copy in Ralph Carlisle Smith Collection, Coronado Room, University of New Mexico Library.

28.   General Leslie R. Groves, statement in U.S. Congress, Senate, Committee Hearings, 79th Congress, Senate Library, Vol. 809, 1946, Special, 52.

"The bomb was not built by theoretical physicists," noted Cyril Smith, only half-jokingly, in 1983. Cyril Smith, "Hierarchy of the Structural Misfit," tape, LAHS.

29.   Margaret Gowing, *Independence and Deterrence: Britain and Atomic Energy, 1945–1952* (London: Macmillan Press, Ltd., 1974). Vol. 1: *Policy Making* and Vol. 2 *Policy Execution* tell of the British-American cooperation and the strange story of Klaus Fuchs. See also H. Montgomery Hyde, *The Atom Bomb Spies* (London: Hamish Hamilton, 1980).

30.   Statement by Leo Szilard, U.S. Congress, Senate, Committee Hearings, 79th Congress, Vol. 809, 1946, Special, 290–98; Leslie R. Groves, *Now It Can Be Told: The Story of the Manhattan Project*, 1, 140.

31.   Dorothy McKibbin, "109 East Palace," in "The Atom and Eve," manuscript in LANL archives, 5. Interview with Dorothy McKibbin, April 22, 1982, Santa Fe.

32.   Quoted in Jeremy Bernstein's profile of I. I. Rabi, *New Yorker*, October 21, 1945, 53; Robert Jungk, "Los Alamos—Life in the Shadow of the Atomic Bomb," typescript, Ralph Carlisle Smith Collection. When Freeman Dyson enrolled in the physics department of Cornell in the fall of 1947, he found himself caught in "endless talk about the Los Alamos days." Freeman Dyson, *Disturbing the Universe* (New York: Harper and Row, 1979), 51.

33. *Los Alamos: Beginning of an Era, 1943–1945* (Los Alamos: LASL, c. 1960), 79–88; Edith C. Truslow, *Manhattan District History: Non-scientific Aspects of Los Alamos Project Y, 1942 through 1946* (Los Alamos: LASL, 1973), LA-5200. The best accounts of life on the Hill, however, are by the wives who shared the experiences: Eleanor Jette, *Inside Box 1633* (Los Alamos: LAHS, 1977); Bernice Brode, "Tales of Los Alamos," *LASL Community News*, June 2–September 22, 1960; "The Atom and Eve," unpublished manuscript, LANL archives; and Marie Kinzel, "The Town of Beginning Again," *Survey Graphic* 35 (October 1946): 354–57 ff. As Marge Schreiber later remarked, "When you're making history, you have no idea you are." "Reminiscences," tape, LAHS.

34. Mario Balifrera, interview with Robert R. Wilson (1979) typescript, LANL. Wilson considered it "a privilege" to have worked at Los Alamos from 1943 to 1945.

35. Interview with Hans Bethe (1975), tape, LAHS. Unnamed physicist quoted in "Prologue," *Bulletin of the Atomic Scientists* 26 (June 1970): 2.

36. *In the Matter of J. Robert Oppenheimer* (Washington, D.C.: Government Printing Office, 1954): 12–13; *The Reminiscences of Kenneth T. Bainbridge*, 42–43, COHP.

37. Alice Kimball Smith, *A Peril and a Hope: The Scientists' Movement in America: 1945–1947* (Chicago: University of Chicago Press, 1965), quoted 5. See also interview with Richard Feynman (by Lillian Hoddeson and George Baym), April 16, 1979, and interview with Robert Serber (by Lillian Hoddeson), March 26, 1979, typescripts, LANL.

38. *Los Alamos: Beginning of an Era*, quoted, 57. On the city of Los Alamos itself, see Marjorie Bell Chambers, "Technically Sweet Los Alamos: The Development of a Federally Sponsored Scientific Community" (Ph.D. dissertation, University of New Mexico, 1974); and James W. Kunetka, *City of Fire: Los Alamos and the Atomic Age, 1943–1945* (Albuquerque: University of New Mexico Press, revised ed., 1979).

39. Groueff, *Manhattan Project*, 257. On Oppenheimer, see Peter Goodchild, *J. Robert Oppenheimer: Shatterer of Worlds* (Boston: Houghton Mifflin Company, 1981) and the collection of letters edited by Alice Kimball Smith and Charles Werner, *Robert Oppenheimer: Letters and Recollections* (Cambridge, Mass.: Harvard University Press, 1980). A good short biography may be found in the series of articles on his death in *Physics Today* 20 (October 1967): 35–53. The latest study is James W. Kunetka, *Oppenheimer: The Years of Risk* (Englewood Cliffs: Prentice-Hall, 1982). The Bradbury quotation is from his "Reminiscences," tape, LAHS.

40. Aage Bohr, "The War Years and the Prospects Raised by the Atomic Weapons," in S. Rozenthal, ed., *Niels Bohr* (Amsterdam: North-

Holland Publishing Company, 1967), 191–214. *Physics Today* 16 (October 1963) is devoted primarily to Niels Bohr's career.

41. S. K. Allison, "A Tribute to Enrico Fermi," *Physics Today* 8 (January 1955): 9; Laura Fermi, *Atoms in the Family: My Life with Enrico Fermi* (Chicago: University of Chicago Press, 1954) is a lively account by his wife; Emilio Segrè, *Enrico Fermi: Physicist* (Chicago: University of Chicago Press, 1970); Lawrence Badash, "Enrico Fermi and the Development of Nuclear Energy," in Carroll W. Pursell, Jr., *Technology in America* (Cambridge, Mass.: MIT Press, 1981), 213–27. *Reviews of Modern Physics* 27 (July 1975): 249–72 presents the papers from a memorial symposium held in Fermi's honor.

42. Victor F. Weisskopf, *Physics in the Twentieth Century: Selected Essays* (Cambridge, Mass.: MIT Press, 1972).

43. S. M. Ulam, *Adventures of a Mathematician.*

44. Stanley A. Blumberg and Gwinn Owens, *Energy and Conflict: The Life and Times of Edward Teller* (New York: G. P. Putnam's Sons, 1976); Edward Teller, *Energy from Heaven and Earth* (San Francisco: W. H. Freeman, 1979), 137–55; Edward Teller (with Allen Brown), *The Legacy of Hiroshima* (Garden City, N.Y.: Doubleday and Company, 1962); Eugene P. Wigner, "An Appreciation on the 60th Birthday of Edward Teller," 1–6, in Hans Mark and Sidney Fernbach, eds., *Properties of Matter under Unusual Conditions* (New York: John Wiley & Sons, 1969).

45. See Jeremy Bernstein's "Profile," in the *New Yorker*, October 13, 1975 and October 21, 1975; and I. I. Rabi, *Science: The Center of Culture.*

46. S. Ulam et al., "John von Neumann, 1903–1957," in Fleming and Bailyn, *The Intellectual Migration*, 235–69.

47. Jeremy Bernstein, *Hans Bethe: Prophet of Energy* (New York: Basic Books, 1980).

48. Otto R. Frisch, *What Little I Remember* (Cambridge: Cambridge University Press, 1979), 151. Otto R. Frisch, "Reminiscences," tape, LAHS.

49. Jane Wilson, "Not Quite Eden," in "The Atom and Eve," 12, LANL.

50. I. I. Rabi, "Can Mankind Survive Its Powers?" (The World Since World War II), 3rd Annual Oppenheimer Memorial Lecture, tape, LAHS. Groves, as quoted in Laura Fermi, *Atoms in the Family*, 226; interview with George Marchi, November 4, 1981, Albuquerque; Elsie McMillan, "Outside the Inner Fence," in *Reminiscences of Los Alamos, 1943–1945*, 43.

51. Charlotte Serber, "Labor Pains," in "The Atom and Eve," 12, LANL.

52. *The Reminiscences of Stafford L. Warren*, 964, UCLA OHP.

53. Olive Kimball Smith and Charles Werner, eds., *Robert Oppenheimer: Letters and Recollections*, 263.

54. Jane Wilson, "Not Quite Eden," in "The Atom and Eve," 12 ff, LANL.

55. Albuquerque *Tribune*, November 27, 1982. Noel Pharr Davis, *Lawrence and Oppenheimer* (New York: Simon and Schuster, 1968) tells of one such falling out.

56. Kathleen Mark, "A Roof over Our Heads," in "The Atom and Eve," LANL; Dan Wilkes, "The Story of the Los Alamos 'Campus,'" *California Monthly* (June 1950), pamphlet, copy Ralph Carlisle Smith Collection. See also the irenic comments by Norris Bradbury in his "Reminiscences," tape, LAHS, and even those by Hans Bethe in his otherwise critical "Comments on the History of the H-Bomb," *Los Alamos Science* 3 (Fall 1982): 44, 49.

57. "Personalities around the Hill: Dr. David Hawkins," *Los Alamos Times* (July 26, 1946): 2; *Manhattan District History, Los Alamos Project Y*, University Publications of America, Microfilm, Reel 12: 7.8–7.10.

58. Charlotte Serber, "Labor Pains," in "The Atom and Eve," LANL; Robert Jungk, "Los Alamos—Life in the Shadow of the Atomic Bomb," typescript, Ralph Carlisle Smith Collection.

59. Theodore F. Koop, *Weapon of Silence* (Chicago: University of Chicago Press, 1946) details the elaborate censorship program.

60. Interview with Lilli Marjon, September 2, 1981, Albuquerque, New Mexico.

61. Interview with Marvin and Ruby Wilkening, January 12, 1982, Socorro, New Mexico; Ruth Marshak, "Secret City" in "The Atom and Eve," 2, LANL; Bernice Brode, "Tales of Los Alamos," *LASL Community News* (June 2, 1960): 4.

62. William L. Laurence, "Address" in *The Beginnings of the Nuclear Age* (New York: Newcomen Society in North America, 1969), 15–18.

63. New York University physics professor Irving S. Lowen wrote to Roosevelt numerous times to voice his concern. See James Conant to Roosevelt, November 18, 1943, Atomic Bomb folder, FDR.

64. Harvey H. Bundy, "Remembered Words," *Atlantic Monthly* 199 (March 1957): 57.

65. Kenneth T. Bainbridge, "Prelude to Trinity," *Bulletin of the Atomic Scientists* 31 (April 1975): 43.

66. David Irving, *The German Atomic Bomb: The History of Nuclear Research in Nazi Germany* (New York: Simon and Schuster, 1967)

and Alan D. Beyerchen, *Scientists under Hitler* (New Haven: Yale University Press, 1977) are the two best studies.

67. Marjorie Bell Chambers, "Technically Sweet Los Alamos," 121–23.

68. W. Heisenberg, "Research in Germany on the Technical Application of Atomic Energy," *Nature* 160 (August 16, 1947): 211–15.

69. Samuel A. Goudsmit, *Alsos* (New York: Henry Schuman, 1947); Boris T. Pash, *The Alsos Mission* (New York: Award House, 1969). See also Michael Bar-Zohar, *The Hunt for German Scientists* (New York: Hawthorne Books, 1967).

70. Quoted in Martin J. Sherwin, *A World Destroyed: The Atomic Bomb and the Grand Alliance* (New York: Alfred A. Knopf, 1975), 145.

71. Gordon Thomas and Max Morgan Witts, *Enola Gay* (New York: Pocket Books, 1978), 201–2.

72. John H. Manley, "A New Laboratory Is Born," in Badash, Hirschfelder, and Broida, eds., *Reminiscences of Los Alamos, 1943–1945*, 33–34.

73. George B. Kistiakowsky, "Reminiscences of Wartime Los Alamos," ibid., 56.

Chapter Two

1. Kenneth T. Bainbridge, *Trinity* (Los Alamos: LASL, 1976), LA-6400-H. David Hawkins, *Manhattan District History, Project Y, The Los Alamos Project* (Los Alamos: LASL, 1961), LAMS-2432 (Vol. 1). This was reissued in 1983 as part of Los Alamos's fortieth anniversary celebration. Interview with Robert W. Henderson, August 24, 1982, Albuquerque, New Mexico.

2. Interview with Robert W. Henderson, August 24, 1982, Albuquerque, New Mexico.

3. David Hawkins, *Manhattan District History, Project Y, The Los Alamos Project*, 267.

4. Herbert E. Bolton, *Coronado on the Turquoise Trail: Knight of Pueblos and Plains* (Albuquerque: University of New Mexico Press, 1949), 311; George P. Hammond and Agapito Rey, *Don Juan de Oñate: Colonizer of New Mexico, 1595–1628* (Albuquerque: University of New Mexico Press, 1953), 317.

5. Marc Simmons, "Full Story of Bloody Jornada del Muerto Would Take a Book," *New Mexico Independent* 86 (April 2, 1982): 2.

6. Hammond and Rey, *Oñate*, 317; J. Manuel Espinosa, trans., *First Expedition of Vargas into New Mexico, 1692* (Albuquerque: University of New Mexico Press, 1940), 61–64.

7. Josiah Gregg, *Commerce of the Prairies* (Norman: University

of Oklahoma Press, 1954); John Fox Hammond, *A Surgeon's Report on Socorro, N.M.: 1852* (Santa Fe: Stage Coach Press, 1966), 21.

8. Albuquerque *Journal* (December 10, 1941).

9. Henry L. Stimson to John Dempsey, January 8, 1943; John Dempsey to Henry L. Stimson, January 27, 1943. John Dempsey papers, Folder "Military Bases," NMSRA.

10. Kit Miniclier, "A-Bomb Site Landowners 'Aged Out,'" *Denver Post* (March 8, 1982).

11. Albuquerque *Tribune* (October 16, 1982); Albuquerque *Journal* (October 17, 1982); Fritz Thompson, "Behind the Barbed Wire Barricade," *Impact*, Albuquerque *Journal* Magazine (October 16, 1982), 4–9.

12. Typescripts, "Selection of Test Site," "Test Site," "Section X-2D, Selection of Test Site,"MED, Trinity, Box 28, LANL.

13. K. T. Bainbridge, "Test Site," typescript in "Trinity Construction Site," LANL.

14. Kenneth T. Bainbridge, "Prelude to Trinity," *Bulletin of the Atomic Scientists* 31 (April 1975): 46.

15. Conversation with Sam P. Davalos, June 29, 1981 (telephone); Lenore Fine and Jesse A. Remington, *United States Army in World War II: The Technical Services: The Corps of Engineers: Construction in the United States* (Washington, D.C.: Government Printing Office, 1972), 699–70; "Construction and Equipment Requirements for Proposed Test Site, Trinity," October 10, 1944, MED, Trinity Planning, Box 28, LANL.

16. John H. Williams to Capt. S. P. Davalos, March 16, 1945, MED, Trinity Planning, Box 28, LANL; J. DeWire to Lt. Comdr. T. M. Keiller, June 11, 1945, MED, Trinity Planning, Box 28, LANL; J. H. Williams to Capt. S. P. Davalos, April 17, 1945, MED, Trinity, Box 28, LANL.

17. Subject: Construction Progress—Trinity, December 7, 1944, MED, Trinity, Box 28, LANL.

18. Construction and Equipment Requirements for Proposed Test Site, Trinity, MED, Trinity Construction, LANL.

19. Kenneth T. Bainbridge, "A Foul and Awesome Display," *Bulletin of Atomic Scientists* 31 (May 1975): 40.

20. Fine and Remington, *The Corps of Engineers Construction in the United States,* 71; John H. Williams to S. P. Davalos, March 31, 1945, MED, Trinity Construction, LANL. Interview with Ted Brown, June 25, 1982, Albuquerque, New Mexico.

21. Interview with Sam P. Davalos, July 12, 1982, Santa Fe, New Mexico.

22. Interview with Ted Brown, June 25, 1982.

23. Mike Gallagher, "Builder of 'Ground Zero' Recalls Crucial Project," Albuquerque *Journal*, December 15, 1980.

24. K. T. Bainbridge to Col. G. B. Tyler, May 23, 1945, MED, Trinity, Box 28, LANL.

25. K. T. Bainbridge to R. W. Henderson, March 15, 1945, MED, Trinity, Box 28, LANL; G. R. Tyler, Subject: Additional Work, Trinity, December 15, 1944, MED, Trinity, Box 28, LANL.

26. Interview with Roy W. Carlson (telephone), September 19, 1982.

27. Lenore Dils, *Horny Toad Man* (El Paso: Boots and Saddle Press, 1966), 167–69.

28. Mario Balibrera interview with George Kistiakowsky (1981), typescript, LANL.

29. Bainbridge, "A Foul and Awesome Display," 44.

30. Interview with Dorothy McKibbin, April 22, 1982, Santa Fe.

31. Interview with J. Hubbard, September 3, 4, 1982, San Jose, Ca.

32. Minutes of a meeting held March 8, 1945 on Trinity conditions, MED, Trinity, Box 28, LANL; The Reminiscences of John Harold Anderson, tape, copy lent by J. Hubbard.

33. Interview with J. Hubbard, September 4, 1982.

34. Kenneth T. Bainbridge, *Trinity*.

35. Ibid. See also G. B. Kistiakowsky to J. R. Oppenheimer, "Activities at Trinity," October 13, 1944, MED, Trinity, Box 28, LANL.

36. J. R. Oppenheimer to All Group Leaders, June 14, 1945, MED, Trinity Planning, Box 27, LANL. L. Fussell, Jr. to K. T. Bainbridge, December 23, 1944, MED, Trinity Planning, Box 27, LANL.

37. Bainbridge, "A Foul and Awesome Display," 40–46.

38. Marjorie Bell Chambers, "Technically Sweet Los Alamos: The Development of a Federally Sponsored Scientific Community" (Ph.D. dissertation, University of New Mexico, 1974); "Trinity Experiments for July 1945 Test," MED, Trinity Planning, Box 27, LANL.

39. K. T. Bainbridge, "Proposed Rehearsal Schedule," June 9, 1945, MED, Trinity Planning, Box 27, LANL.

40. Kenneth T. Bainbridge, *Trinity*.

41. *The Atom* 11 (May–June 1980): 14–15.

42. Interview with Robert W. Henderson.

43. Interview with John Magee (telephone), August 30, 1982; interview with Jack Hubbard, September 3, 4, 1982; interview with Jerry Jercinovic, August 29, 1982, Albuquerque.

44. I. I. Rabi, "How Well We Meant," lecture, Los Alamos, April 14, 1983, tape, LAHS.

45. Robert Jungk, *Brighter Than a Thousand Suns: A Personal History of the Atomic Scientists* (New York: Harcourt Brace Jovanovich, 1958), 197.

46. The Reminiscences of Stafford L. Warren, 770–71, 775, UCLA OHP.

47. Interview with Sam P. Davalos, July 12, 1982, Santa Fe.

48. Interview with Robert W. Henderson.

49. Lansing Lamont, *Day of Trinity* (New York: Atheneum, 1965), 70.

50. Chambers, "Technically Sweet Los Alamos."

51. K. T. Bainbridge to J. R. Oppenheimer, March 15, 1945, MED, Trinity, Box 27, LANL.

Chapter Three

1. Ferenc M. Szasz, *The Divided Mind of Protestant America, 1880–1930* (Tuscaloosa: University of Alabama Press, 1982) discusses this.

2. Quoted in Ronald W. Clark, *The Scientific Breakthrough* (New York: G. P. Putnam's Sons), 152.

3. F. W. Aston, *Isotopes* (London, 1922), in Kenneth T. Bainbridge, "A Foul and Awesome Display," *Bulletin of Atomic Scientists* 31 (May 1975): 44; William L. Laurence, *The Hell Bomb* (New York: Alfred A. Knopf, 1951), 5–6.

4. Edward Teller, "The Work of Many People," *Science*, 121 (February 1955): 267–72 gives a good account of this.

5. Peter Goodchild, *J. Robert Oppenheimer, Shatterer of Worlds* (Boston: Houghton Mifflin Company, 1981), 52–54.

6. Pearl S. Buck, "The Bomb—The End of the World," *American Weekly* (March 9, 1959): 9–12; H. D. Dudley, "The Ultimate Catastrophe," *Bulletin of the Atomic Scientists* 31 (November 1975): 22.

7. Bainbridge, "A Foul and Awesome Display," 44.

8. Interview with Hans Bethe, November 11, 1981, Albuquerque, New Mexico; Edward Teller to author, December 15, 1981.

9. Interview with John Magee (telephone), April 15, 1982.

10. Teller and Konopinsky, December 2, 1943, Report at Los Alamos. This is still classified, but one page was cleared January 14, 1982, at the author's request.

11. *New York Times* (November 30, 1975) 50: 4.

12. Santa Fe *New Mexican* (November 27, 1945).

13. El Paso *Times* (August 8, 1945).

14. Hans A. Bethe, "Can Air or Water Be Exploded?" *Bulletin of the Atomic Scientists* 1 (March 15, 1946): 2, 14.

15. Statement of Hans Bethe. U.S. Congress, Senate, Committee Hearings, 79th Congress, Vol. 809, 1946, Special, 224–25.

16. *New York Times* (November 30, 1975) 50: 4.

17. Jonathan Schell, *The Fate of the Earth* (New York: Alfred A. Knopf, 1982), 96; Cincinnati *Enquirer* (November 27, 1975); *Time* (November 14, 1983): 43; *New Yorker* (November 21, 1983): 41; *Parade Magazine* (October 30, 1983): 4–7.

18. Albuquerque *Journal* (August 26, 1945) in War Records Library Collection, Scrapbook 71, New Mexico State Record Center and Archives, Santa Fe; notes from interview with Wright Langham, Lansing Lamont Papers, HST.

19. Albuquerque *Journal* (July 12, 1970); notes from interview with Berlyn Brixner and Richard J. Watts, Lansing Lamont Papers, HST.

20. Bainbridge, "A Foul and Awesome Display," 44; Leslie R. Groves, *Now It Can Be Told: The Story of the Manhattan Project* (London: André Deutsch, 1963), 296–97; *Los Alamos: Beginning of an Era, 1943–1945.*

21. Ruth Marshall, "Secret City," in "The Atom and Eve," manuscript, LANL, 15.

22. James Cornell, *The Great International Disaster Book* (New York: Pocket Books, 1976), 255–57.

23. K. T. Bainbridge to Capt. Thomas O. Jones, April 19, 1945, 322, Trinity Measurements, MED, Oppenheimer Collection, Box 13, LANL.

24. L. Don Leet to J. Robert Oppenheimer, June 26, 1945: Subject: Ground Vibrations from Trinity Tests. Oppenheimer Collection, MED, Box 13, File no. 322: Trinity Measurements, LANL.

25. Ibid.

26. L. Don Leet to J. R. Oppenheimer, September 7, 1945, Oppenheimer File, LANL. Santa Fe *New Mexican* (November 8, 1945).

27. L. Don Leet, "Earth Motion from the Atomic Bomb Test," *American Scientist* 34 (April 1946): 198–211.

28. Gretchen Heitzler, *Meanwhile, Back at the Ranch* (Albuquerque: Hidden Valley Press, 1980), 126.

29. Interview with John Magee, April 15, 1982 (telephone).

30. Joseph O. Hirschfelder, "The Scientific and Technological Miracle at Los Alamos," in Laurence Badash, Joseph O. Hirschfelder, and Herbert P. Broida, eds., *Reminiscences of Los Alamos, 1943–1945* (Dordrecht, Holland: D. Reidel Publishing Company, 1980), 73–76.

31. H. L. Anderson, *Radioactive Measurements at the 100 Ton Trial* (May 25, 1945), LA-282.

32. J. O. Hirschfelder and John Magee to K. T. Bainbridge, June 16, 1945, LANL and LA-1027-DEL.

33. J. O. Hirschfelder and John Magee to K. T. Bainbridge, July 6, 1945, 322 Trinity Measurements, MED, Oppenheimer, Box 13, LANL; interview with John Magee, August 30, 1982 (telephone).

34. L. H. Hempelmann, "Hazards of 100 Ton Shot at Trinity," May

18, 1945, in L. H. Hempelmann, "Preparation and Operational Plan of Medical Group (TR-7) for Nuclear Explosion, July 16, 1945," LA-631; H. L. Anderson to K. T. Bainbridge, April 19, 1945, MED, Trinity, 322, Box 6, LANL.

35.   S. K. Allison to J. R. Oppenheimer, May 10, 1945, 322—Trinity Measurements, MED, Oppenheimer, Box 13, LANL.

36.   Lansing Lamont, *Day of Trinity* (New York: Atheneum, 1965), 127. *The Reminiscences of Stafford L. Warren,* 783, UCLA OHP. Interview with James F. Nolan, May 25, 1983 (telephone).

37.   *The Reminiscences of Stafford L. Warren,* 799.

38.   Interview with John Magee.

39.   Laura Fermi, *Illustrious Immigrants: The Intellectual Migration from Europe, 1930–1941* (Chicago: University of Chicago Press, 1968), 192–93; William Laurence, in the *New York Times* (June 29, 1951) 4: 2.

40.   Interview with Stanislaw M. Ulam, July 27, 1983, Santa Fe.

41.   *New York Times* (June 29, 1951), 4: 2.

Chapter Four

1.   Clark C. Spence, "The Cloud Crackers: Moments in the History of Rainmaking," *Journal of the West* 18 (October 1979): 69.

2.   *Journal of J. Hubbard, Meteorologist;* copies lie in the archives of Caltech and Los Alamos, and in the author's possession.

3.   Jack M. Hubbard, *100-ton Test Meteorological Report,* LA-285.

4.   Hubbard, *100-ton Test,* 2.

5.   Henry Lewis Stimson Diaries, 51: 159–60, Microfilm edition, reel 9, Manuscripts and Archives, Yale University Library, New Haven, Connecticut.

6.   Hubbard, *Journal,* 81.

7.   Jack M. Hubbard, *Nuclear Explosion: Meteorological Report* (September 10, 1945), LA-357.

8.   Hubbard, *Journal,* 77.

9.   Interview with Jack M. Hubbard, November 29, 1981; September 3, 4, 1982.

10.   Hubbard, *Nuclear Explosion,* 6.

11.   Ibid., 7.

12.   Ibid.

13.   *The Reminiscences of Stafford L. Warren,* 784, 800–801, UCLA OHP.

14.   Hubbard, *Journal,* 83–115; Jack M. Hubbard, *Reminiscences,* tape in author's possession.

15.   Hubbard, *Journal,* 90.

16. Hubbard, *Nuclear Explosion*, 29.

17. Interview with Sam P. Davalos, July 12, 1982, Santa Fe; interview with Leo M. Jercinovic, August 29, 1982, Albuquerque.

18. Hubbard, Weather Forecast July 15, 1945, 322 Trinity Measurements, MED, Oppenheimer, Box 13, LANL.

19. Los Alamos *Monitor* (April 4, 1983).

20. Leslie R. Groves, "Some Recollections of July 16, 1945," *Bulletin of the Atomic Scientists* 26 (June 1970): 26.

21. Interviews with John Magee, April 15, 1982; August 30, 1982. See also William L. Laurence, *Dawn Over Zero: The Story of the Atomic Bomb* (New York: Alfred A. Knopf, 1946), 189–92.

22. J. O. Hirschfelder to J. R. Oppenheimer, April 25, 1945 in LA-1027-DEL.

23. Hubbard, *Journal*, 98. Current meteorological theory holds that clouds tend to rise to an equilibrium temperature and that momentum is not much of a consideration; interview with Jack Reed, January 20, 1982, Albuquerque.

24. Interview with Jack M. Hubbard (telephone), November 29, 1981; S. K. Allison to J. R. Oppenheimer, May 10, 1945, 322, Trinity Measurements, MED, Oppenheimer, Box 13, LANL.

25. Los Angeles *Herald Examiner* (July 10, 1983); Albuquerque *Tribune* (July 11, 1983); interviews with John Magee, April 15, 1982; August 30, 1982 (telephone).

26. Hubbard, *Journal*, 104.

27. Ibid., 108–11. Twenty-five years later, when Groves recalled the weather of July 16, 1945, he had moderated his feelings only slightly. See Leslie R. Groves, "Some Recollections of July 16, 1945," *Bulletin of the Atomic Scientists* 26 (June 1970): 24–26.

28. G. R. Tyler, "Notes on the Manhattan Engineer District Activities, New Mexico—1944–45," 15, typescript in author's possession.

29. Groves, "Some Recollections of July 16, 1945," 25.

Chapter Five

1. *Los Alamos, 1943–1945: Beginning of an Era* (Los Alamos: LASL, c. 1960), 43.

2. Elsie McMillan, "Outside the Inner Fence," in Lawrence Badash, Joseph O. Hirschfelder, and Herbert P. Broida, eds., *Reminiscences of Los Alamos, 1943–1945* (Dordrecht, Holland: D. Reidel Publishing Company, 1980), 45.

3. *The Reminiscences of Kenneth T. Bainbridge*, 56–57, COHP.

4. Interview with Norris Bradbury, July 30, 1982, Los Alamos.

5. James W. Kunetka, *City of Fire: Los Alamos and the Atomic*

Age, 1943–1945 (Albuquerque: University of New Mexico Press, 1979; rev. ed.), 160.

6. David Hawkins, *Manhattan District History Project Y: The Los Alamos Project*. Vol. 1: *Inception until August, 1945*, LAMS-2432, 274; Robert Cahn, "Behind the First Atomic Bomb," *Saturday Evening Post* (July 16, 1960): 73. See also Cyril S. Smith, "Metallurgy at Los Alamos, 1943–1945," *Metal Progress* 65 (May 1954): 87. Boyce McDaniel, "Journeyman Physicist," in Jane Wilson, ed., *All in Our Time: The Reminiscences of Twelve Nuclear Pioneers* (Chicago: Bulletin of the Atomic Scientists, 1975), 174–88. Mario Balibera, interview with Robert Bacher (1980), typescript, LANL.

7. Bradbury, in Santa Fe *New Mexican* (July 16, 1965).

8. Interview with John Magee, March 30, 1982 (telephone).

9. George B. Kistiakowsky, "Trinity—A Reminiscence." *The Bulletin of the Atomic Scientists* 36 (June 1980): 20. William L. Laurence, "Lightning Blew Up Dummy Atom Bomb," *New York Times* (September 27, 1945).

10. George Fitzpatrick, "Atomic Bomb Sight-Seeing," *New Mexico* 24 (January 1946): 35.

11. The San Diego *Union* (July 11, 1965).

12. Mario Balibrera, interview with Victor F. Weisskopf, 1981, typescript, LANL.

13. Richard C. Hindley, "The Day the Sun Came Out of the Sky," *PGH Press* (July 11, 1976); Charles A. Thomas, "Epoch in the Desert," *Monsanto Magazine* 24 (1945): 28–31.

14. A concise account of the blast can be found in Richard G. Hewlett and Oscar E. Anderson, Jr., *The New World, 1939–1946: A History of the United States Atomic Energy Commission*, Vol. 1 (University Park, Pa.: Penn State University Press, 1962), 376–80; Edwin C. Bearss, Trinity Test Site Interview [with Krohn] typescript, June 1968, 45. Folder, Trinity Site, Raw Data, NPS.

15. Kunetka, *City of Fire*, 168–69.

16. J. E. Mack, "Weapon Data Effects and Instruments," LA-1024 (LASL, 1947), 34–41. Stafford L. Warren, *The Role of Radiology in the Development of the Atomic Bomb*, a reprint of a chapter from *Radiology in World War II*, Kenneth D. A. Allen, ed. (Washington, D.C.: Surgeon General's Office, 1966), 884. Warren always insisted the cloud rose to 70,000 feet. He measured the height of the door jamb with a meter stick and then triangulated the top of the cloud at this distance. No one later agreed with his estimate, but Warren was the only one who actually measured the cloud, and no one had any contrary evidence. *The Reminiscences of Stafford L. Warren*, UCLA OHP, 504.

17. Emilio Segrè, *Enrico Fermi: Physicist* (Chicago: University of

Chicago Press, 1970). There is some debate whether Fermi selected 10,000 or 20,000 tons.

18. Interview with Holm Bursum, June 1, 1981, Socorro, N.M.

19. Tucumcari *Daily News* (July 17, 1945); interview with Thomas Treat, February 22, 1982, Albuquerque, N.M.; El Paso *Herald Post* (July 16, 1945); Las Cruces *Citizen* (August 9, 1945); Artesia *Advocate* (August 9, 1945); Silver City *Press* (July 17, 1945); Carrizozo *Outlook* (July 20, 1945). Elvis E. Fleming, unpublished address to the Roswell Public Library, July 17, 1983. Copy in author's possession.

20. Joseph O. Hirschfelder, "The Scientific and Technological Miracle at Los Alamos," in Badash, Hirschfelder, and Broida, *Reminiscences of Los Alamos, 1943–1945*, 77.

21. El Paso *Herald Post* (July 16, 1945).

22. Steve Lowell, "Pilot Here Still Recalls Vividly the Day 'The Bomb' Was Born," Albuquerque *Journal* (July 16, 1950).

23. Albuquerque *Journal* (July 17, 1945).

24. See, e.g., Silver City *Daily Press* (July 16, 1945); Las Cruces *Sun-News* (July 16, 1945). Copies of contingency releases A, B, C, and D, supplied by Modern Military Branch, Military Archives Division, National Archives (no record group listed).

25. Santa Fe *New Mexican* (March 16, 1951).

26. Lamont, *Day of Trinity*, 244–56.

27. Interview with Millard Hensley, in Lansing Lamont papers, HST.

28. Kit Miniclier, "White Sands a Desolate Historic Site," Denver *Post* (March 10, 1981).

29. Albuquerque *Journal* (August 7, 1945).

30. Stan Green, "'Fat Man' Pops in Big Way in New Mexico Soil," Alamogordo *News* (n.d.), copy in author's possession; Alamogordo *News* (July 19, 1945).

31. Lincoln County *News* (August 10, 1945); Albuquerque *Journal* (August 7, 1945), in War Records Library Collection, Scrapbook 71, New Mexico State Records Center and Archives, Santa Fe.

32. Albuquerque *Journal* (July 12, 1970); interview with McAllister Hull, Jr., December 17, 1982, Albuquerque; Trinity Test Site Interview, recorded by Edwin C. Bearss, June 1968, typescript, folder Trinity Site, Raw Data, NPS, Santa Fe; To: Lt. Taylor from Capt. R. A. Larkin, July 27, 1945, in LANL Photo Archives Historical Documents File; Jeremy Bernstein, Profile of I. I. Rabi, *The New Yorker* (October 21, 1945): 54.

33. *New York Times* (August 10, 1945) 3: 6.

34. Interview with Leo M. (Jerry) Jercinovic, August 29, 1982; interview with Roy W. Carlson, September 19, 1982 (telephone); Jack M. Hubbard, *Journal*, 116.

35. Otto Frisch, *What Little I Remember* (Cambridge: Cambridge University Press, 1979), 164.

36. This report is printed in the Appendix of Leslie R. Groves, *Now It Can Be Told: The Story of the Manhattan Project* (London: André Deutsch, 1963).

37. It is printed, among other places, in the *New York Times* (July 16, 1970) 20: 6.

38. Hubbard, *Journal*, 123.

39. Carlsbad Daily *Current-Argus* (August 17, 1945).

40. Kunetka, *City of Fire*, 170.

41. William L. Laurence, "Drama of the Atomic Bomb Found Climax in July 16 Test," *New York Times* (September 26, 1945); interview with Dorothy McKibbin, April 22, 1982, Santa Fe.

42. Jeremy Bernstein, Profile of I. I. Rabi, Part 2, *The New Yorker* (October 21, 1975): 58.

43. Notes from interviews with Richard Watts and Berlyn Brixner, Lansing Lamont papers, HST.

44. Kit Miniclier, "N-less Alternate Feared," Denver *Post* (March 9, 1981).

45. Interview with Marvin Wilkening, Socorro.

46. "Thoughts by E. O. Laurence," in *Foreign Relations of the United States, Diplomatic Papers: The Conference of Berlin (the Potsdam Conference)*, 1945, Vol. 2 of 2 [Enclosure 4], 1369–70.

47. K. T. Bainbridge, "A Foul and Awesome Display," *The Bulletin of the Atomic Scientists* 31 (May 1975): 46.

48. "Thoughts by E. O. Lawrence," 1369–70.

49. Albuquerque *Journal* (May 27, 1945).

50. Laura Fermi, "Bombs or Reactors," *Bulletin of the Atomic Scientists* 26 (June 1970): 27.

51. Bernice M. Morgan, "Tuesday Is Anniversary of Trinity Explosion," *Los Alamos Times* (July 12, 1946). Robert Porton, "Talk to Docents," taped reminiscences, LAHS.

52. Albuquerque *Journal* (July 12, 1970).

Chapter Six

1. *The Reminiscences of Stafford L. Warren*, 807, UCLA OHP; interview with Louis Hempelmann, December 22, 1981, Santa Fe, N.M.

2. J. M. Blair, D. H. Frisch, S. Katcoff [and J. M. Hubbard], *Detection of Nuclear Explosion Dust in the Atmosphere*, LA-418.

3. Paul Aebersold, *July 16th Nuclear Explosion—Safety and Monitoring of Personnel*, LA-616, 6–8.

4.   Interview with Herbert L. Anderson, March 16, 1983, near Los Alamos.

5.   Carl Maag and Steve Rohrer, *Project Trinity* (Washington, D.C.: Defense Nuclear Agency, 1983) shows the dosage levels.

6.   Paul Aebersold, *July 16th Nuclear Explosion—Safety and Monitoring of Personnel*, LA-616, 21–23; 33–35; 89; 156–57.

7.   Victor F. Weisskopf, "On Avoiding Nuclear Holocaust," *Technology Review* 83 (October 1980): 28–35.

8.   K. T. Bainbridge to Capt. T. O. Jones, "Legal Aspects of T. R. Tests," May 2, 1945, 310.1, Trinity Planning, MED, Trinity, Box 27, LANL; V. Weisskopf, J. Hoffman, P. Aebersold, and L. Hempelmann to G. Kistiakowsky, September 5, 1945, 322 Trinity Measurements, MED, Oppenheimer, Box 13, LANL.

9.   Victor F. Weisskopf, *Physics in the Twentieth Century: Selected Essays* (Cambridge, Mass.: MIT Press, 1972), 16.

10.   *Manhattan District History Los Alamos Project Y*, 7, University Publications of America Microfilm, Reel 3, 2.10–2.11. A Roentgen is defined as a quantity of X-ray of gamma radiation such that the associated corpuscular emission, for one cubic centimeter of air, produces in that air ions carrying one electrostatic unit of electricity of either sign. A REM, or "Roentgen equivalent in man," is defined as that amount of radiation which produces the same biologic effect as one Roentgen of hard X-rays. Warren, *The Role of Radiology*, 850.

11.   Stafford L. Warren, *The Role of Radiology in the Development of the Atomic Bomb*, a reprint of a chapter from Kenneth A. A. Allen, ed., *Radiology in World II* (Washington, D.C.: Surgeon General's Office, 1966), 833.

12.   William D. Sharpe, "The New Jersey Radium Dial Painters: A Classic in Occupational Carcinogenesis," *Bulletin of the History of Medicine* 52 (Winter 1978): 560–70; Robley D. Evans, "Radium Poisoning: A Review of Present Knowledge," *American Journal of Public Health* 23 (October 1933): 1017–23.

13.   "The Plutonium Project," *Radiology* 49 (September 1947): 269–365; Robert S. Stone, ed., *Industrial Medicine on the Plutonium Project: Survey and Collected Papers* (New York: McGraw-Hill, 1951). Barton C. Hacker, "The Chicago Health Division in the Manhattan Project," Chapter 2 (draft) of manuscript, "Elements of Controversy: A History of Radiation Safety in the Nuclear Weapons Testing Program." Copy lent by author.

14.   Henry A. Blair, ed., *Biological Effects of External Radiation* (New York: McGraw-Hill, 1954), details this effort. See also *New York Times* (December 10, 1945) 2: 6; Carl Voegtlin and Harold Hodge, eds.,

Pharmacology and Toxicology of Uranium Compounds (New York: McGraw-Hill, 1949).

15. Ronald L. Kathren and Paul L. Ziemer, "Introduction: The First Fifty Years of Radiation Protection: A Brief Sketch," in Kathren and Ziemer, eds., *Health Physics: A Backward Glance* (New York: Pergamon Press, 1980), 1–4.

16. Louis H. Hempelmann, *History of the Health Group,* (A-6) (March 1943–November 1945), 314.7, LANL. Hacker, "Radiological Safety at Los Alamos, 1943–1945," Chapter 3 of "Elements of Controversy."

17. Frederic de Hoffmann, "Pure Science in the Service of Wartime Technology," *Bulletin of the Atomic Scientists* 31 (January 1975): 41–44.

18. Otto Frisch, *What Little I Remember* (Cambridge: Cambridge University Press, 1979), 162.

19. Leslie R. Groves, *Now It Can Be Told: The Story of the Manhattan Project* (London: André Deutsch, 1963), 38.

20. Robert S. Stone, "Health Protection Activities of the Plutonium Project," *Proceedings of the American Philosophical Society* 90 (January 1946): 11–19.

21. Stewart Alsop and Ralph E. Lapp, "The Strange Death of Louis Slotin," *Saturday Evening Post* (March 6, 1954): 25 ff.

22. Interview with Kermit Larsen, April 26, 1982 (telephone).

23. J. O. Hirschfelder and John Magee to K. T. Bainbridge, July 6, 1945, 322 Trinity Measurements, MED, Oppenheimer, Box 13, LANL.

24. Inter-Office Memorandum, Registered Letters, September 26, 1945; Nonregistered Letters (no date), MED 329, LANL.

25. *The Reminiscences of Stafford L. Warren,* 780.

26. Interview with Louis Hempelmann, December 22, 1981.

27. Interview with John Magee, August 30, 1982.

28. T. O. Palmer, "Evacuation Detachment at Trinity," 48–50, in Louis H. Hempelmann, *Preparation and Operational Plan of Medical Group (TR-7) for Nuclear Explosion, 16 July 1945,* LA-631; June 26, 1945, Memorandum to File: Assistance of 8th Service Command in Evacuation Plans, OSRD Files, Record Group 227, Records of the Office of Scientific Research and Development, Modern Military Branch, Military Archives Division, National Archives.

29. "Town Monitoring Crew Final Instructions," July 10, 1945, 322 Trinity Measurements, MED, Oppenheimer, Box 13, LANL. L. H. Hempelmann to J. G. Hoffman, "Procedure to be used by town Monitors," July 10, 1945, in LA-631; "Instructions for Monitors," July 10, 1945, ibid.

30. "Changes and Supplement to Town Monitoring," July 7, 1945, 322 Trinity Measurements, MED, Oppenheimer, Box 13, LANL.

31. Field notes of Capt. H. L. Barnett, in Joseph G. Hoffman, *Nuclear Explosion, 16 July 1945: Health Physics Report on Radioactive Contamination throughout New Mexico Following the Nuclear Explosion,* LA-626; interview with Berlyn Brixner, February 24, 1983, Los Alamos.

32. Trinity Test Site interview, recorded by Edwin C. Bearss, June 1968, NPS, Santa Fe; notes from interview with Richard J. Watts and Berlyn Brixner, Lansing Lamont Papers, HST.

33. Paul Aebersold, *July 16th Nuclear Explosion—Safety and Monitoring of Personnel,* in LA-616.

34. Jack M. Hubbard, *Journal,* 126; *The Reminiscences of Kenneth T. Bainbridge,* COHP.

35. K. T. Bainbridge, ed., *Weapon Data: Effects and Instruments,* LASL Microfiche 1024 (October 27, 1947).

36. Field notes of Carl S. Hornsberger, LAMD-329, n.p.

37. K. T. Bainbridge to author, November 8, 1981.

38. Field notes made by radiation monitor John Magee on July 16, 1945, LAMD-329.

39. Interview with Louis Hempelmann, January 26, 1982. Events in camp immediately following shot (summarized from Col. Warren's and Hempelmann's personal notes), in Louis H. Hempelmann, *Preparation and Operational Plan of Medical Group (TR-7) for Nuclear Explosion, 16 July 1945,* LA-631.

40. Interview with John Magee, August 30, 1982.

41. Joseph G. Hoffman, *Nuclear Explosion, 16 July 1945: Health Physics Report on Radioactive Contamination Throughout New Mexico Following the Nuclear Explosion,* LA-626, 31.

42. Map drawn by Warren (July 1945) in RG 77, Records of the Office of the Chief of Engineers, Manhattan Engineer District, MED TS, folder 4, National Archives.

43. Field notes by Bob Leonard, LA-329.

44. Conference about Contamination of Countryside near Trinity with Radioactive Materials, July 10, 1945 in LA-631.

45. Interview with Hymer Friedell, July 8, 1983 (telephone).

46. Notes from interview with Wright Langham, Lansing Lamont papers, HST.

47. "Conference about Contamination," in LA-631.

48. Santa Fe *New Mexican* (May 17, 1946).

49. Kit Miniclier, "Visit by Soldiers Told Them of Import," Denver *Post* (March 10, 1981); Hempelmann interview, December 22, 1981; January 16, 1982; George Fitzpatrick, "Atomic Bomb Sight-seeing," *New Mexico* 24 (January 1946): 14.

50. Statement of General Leslie R. Groves in U.S. Congress, Senate, Committee Hearings, 79th Congress, Vol. 809, 1946, Special, 36–37.

51. Interview with James F. Nolan (telephone), May 24, 1983.

52. Interview with Richard Watts, tape, LAHS.

53. Interview with Hempelmann, January 26, 1982.

Chapter Seven

1. Stafford L. Warren to Brig. Gen. James McCormack, Jr., November 28, 1947, in *Trinity Survey Program* (1947), UCLA-22. Joseph G. Hoffman, *Nuclear Explosion, 16 July 1945: Health Physics Report on Radioactive Contamination throughout New Mexico Following the Nuclear Explosion*, LA-626, 48–50.

2. Thomas Ewing Dabney, "Red Hair of Hereford Cattle in Region Surrounding Site of Atomic Bomb Test in New Mexico Turns White," *New Mexico Stockman* 10 (November 1945): 41.

3. Interview with Dr. Louis Hempelmann, December 22, 1981; interview with Holm Bursum, June 1, 1982. Wright H. Langham to Madame Jacqueline Juillard, January 11, 1960, in LANL Photo Archives.

4. Clovis *News Journal*, undated clipping, but late November 1945, War Records Library Collection, Scrapbook 71, New Mexico State Records Center and Archives, Santa Fe. Hereafter Scrapbook 71, NMSRA.

5. Albuquerque *Journal* (February 1, 1946), Scrapbook 71, NMSRA.

6. Louis H. Hempelmann, "Transfer of Alamogordo Cattle," June 7, 1948, Record Group 431, LANL Microfilm.

7. Interview with Holm Bursum, June 1, 1982.

8. Albuquerque *Journal* (December 12, 1945), Scrapbook 73, NMSRCA.

9. Las Vegas *Daily Optic* (July 11, 1947); Santa Fe *New Mexican* (June 7, 1946); Albuquerque *Journal* (April 15, 1946), Scrapbook 73, NMSRA.

10. Hempelmann to A. W. Betts, March 13, 1947, Record Group 431, LANL.

11. Thomas L. Shipman to Carroll Tyler, September 5, 1952, Record Group 431, LANL.

12. Albuquerque *Journal* (July 16, 1949).

13. Carroll Tyler, "Transfer of Alamogordo Cattle," June 10, 1948; Wright H. Langham to Robert Buettner, October 14, 1949; Albert W. Bellamy to James H. Jensen, November 4, 1948; Carroll Tyler, "Transfer of Alamogordo Cattle," July 28, 1948; Wright H. Langham to Albert W. Bellamy, March 6, 1949, Record Group 431, LANL.

14. Wright H. Langham to Madame Jacqueline Juillard, January 11, 1960, LANL Photo Archives.

15. Samuel Glasstone and Philip T. Dolan, *The Effects of Nuclear Weapons* (Washington, D.C.: Government Printing Office, 1977, 3rd

ed.), 621–22; Hempelmann interview, December 22, 1981; Wright H. Langham to Major R. T. Veenstra, July 30, 1953, Record Group 431, LANL.

16. Interview with Neil Dilley, retired Eastman Kodak executive, December 23, 1981, Albuquerque.

17. J. H. Webb, "The Fogging of Photographic Film by Radioactive Contaminants in Cardboard Packaging Materials," *Physical Review* 76 (August 1, 1949): 375–80.

18. Ibid., 379–80.

19. *New York Times* (May 23, 1946) 1: 3; (May 26, 1946) 7: 7; Kermit H. Larson, *Continental Close-in Fallout: Its History, Measurement and Characteristics* (typescript report, n.d., copy in author's possession). Larson's report is being printed in Paul Dunaway et al., eds., *Radioecology of Plutonium and Other Radionuclides in Desert Ecosystems* (Report No. NVO 224; U.S. Dept. of Energy, Las Vegas, Nevada Office) (forthcoming).

20. *The Reminiscences of Stafford L. Warren*, 1174, UCLA OHP.

21. Clear Creek, N. M. to Washington Liaison office [Hempelmann to Groves], August 25, 1945, MED, Trinity Measurements, Oppenheimer, Box 13, LANL; Paul Aebersold and P. B. Moon, *July 16th Nuclear Explosion: Radiation Survey of Trinity Site Four Weeks after Explosion*, LA-357; interview with Kermit Larson, May 21, 1982.

22. *Trinity Survey Program* (1947), UCLA-22, 45.

23. K. H. Larson et al. [UCLA-108], *Alpha Activity Due to the 1945 Atomic Bomb Detonation at Trinity, Alamogordo, New Mexico* (1951), 4.

24. UCLA-22, Introduction 1–12, 31.

25. UCLA-22, 54.

26. UCLA-22, 60.

27. Stafford L. Warren to Brig. Gen. James McCormack, Jr., November 28, 1947, in *Trinity Survey Program* (1947), UCLA-22; also 66.

28. UCLA-32, Part 2, 22.

29. *The 1948 Radiological and Biological Survey of Areas in New Mexico Affected by the First Atomic Bomb Detonation* (1949), UCLA-32, 17–22. *The Reminiscences of Stafford L. Warren*, 489, UCLA OHP.

30. K. H. Larson et al. [UCLA-140], *The 1949 and 1950 Radiological Soil Survey of Fission Product Contamination and Some Soil-Plant Interrelationships of Areas in New Mexico Affected by the First Atomic Bomb Detonation* (1951), 37–38.

31. UCLA-108, 40.

32. Interview with Kermit Larson, May 21, 1982.

33. Larson, *Continental Close-In Fallout*, UCLA-140, 20.

34. UCLA-32, Part 2, 40.

35. Both Larson interviews; Larson, *Continental Close-In Fallout*, UCLA-140, 34–39.

36. Interview with Kermit Larson, April 26, 1982; May 22, 1982.

37. Alamogordo *News* (July 21, 1949).

38. UCLA-32, Part 2, 43.

39. UCLA-108, 34, 36.

40. Field notes by Bob Leonard, LA-329.

41. Barton C. Hacker, "Trinity," Chapter 4 (ms) of "Elements of Controversy: A History of Radiation Safety in the Nuclear Weapons Testing Program." Copy lent by author.

42. Itinerary of trip made by Colonel Warren, Captain Whipple, and L. H. Hempelmann on August 12, 1945, LA-329.

43. Map by Warren in RG 77, Records of the Office of the Chief of Engineers, Manhattan Engineer District, MED, TS, Folder 4, "Trinity," National Archives.

44. V. Weisskopf, J. Hoffman, P. Aebersold, and L. Hempelmann to G. Kistiakowsky, September 5, 1945, 322 Trinity, MED, Oppenheimer, Box 13.

45. Conversation with K. Larson, May 21, 1982.

46. Ernest J. Sternglass, "The Death of All Children," *Esquire* (September 1969), 1a–1d; Denver *Post* (July 27, 1969). Ernest J. Sternglass, *Low-level Radiation* (London: Earth Island, 1973), 86–93.

47. See also Thomas H. Saffer and Orville E. Kelly, *Countdown Zero* (New York: G. P. Putnam's Sons, 1982).

48. Carl Maag and Steve Kohrer, *Project Trinity* (Washington, D.C.: Defense Nuclear Agency, 1983).

49. Stafford L. Warren to Leslie R. Groves, July 21, 1945, RG 77, Records of the Office of the Chief of Engineers, Manhattan Engineer District, MED TS, Folder 4, "Trinity," National Archives, Washington, D.C.

50. Interview with Louis Hempelmann, December 22, 1981; *The Reminiscences of Stafford L. Warren*, 811, UCLA OHP.

Chapter Eight

1. Herbert Feis, "The Secret That Traveled to Potsdam," *Foreign Affairs* 38 (January 1960): 300–317. When James Conant returned to Washington after the test, George Harrison welcomed him with "Congratulations, it worked." Conant replied, "Yes, it worked. As to congratulations, I am far from sure—that remains for history to decide." Conant, *My Several Lives: Memoirs of a Social Inventor* (New York: Harper and Row, 1970), 304. The Stimson quotation is from Harvey

H. Bundy, "Remembered Words," *Atlantic Monthly* 199 (March 1957): 57.

2. Henry Lewis Stimson Diaries 52: 31 (Microfilm edition, reel 9), Manuscripts and Archives, Yale University Library, New Haven, Connecticut.

3. Ibid., 52: 33, 9.

4. It is printed in *Foreign Relations of the United States: Diplomtic Papers: The Conference of Berlin (The Potsdam Conference)* (Washington, D.C.: Government Printing Office, 1945), 2: 1361–70.

5. J. Robert Oppenheimer, "Niels Bohr and Atomic Weapons," *New York Review of Books* (December 17, 1966), 7.

6. Aide-Memoir of conversation between the president and the prime minister at Hyde Park, September 18, 1944, Atomic Bomb folder, FDR. Martin J. Sherwin, *A World Destroyed: The Atomic Bomb and the Grand Alliance* (New York: Alfred A. Knopf, 1975), 107; Barton J. Bernstein, "The Quest for Security: American Foreign Policy and International Control of Atomic Energy, 1942–1946," *Journal of American History* 60 (March 1974): 1007; Alice Kimball Smith, "The Elusive Dr. Szilard," *Harper's Magazine* 221 (July 1960): 77–86; Leo Szilard, "A Personal History of the Atomic Bomb," *University of Chicago Roundtable* (September 25, 1945): 14–16.

7. Elting E. Morison, *Turmoil and Tradition: A Study of the Life and Times of Henry L. Stimson* (Boston: Houghton Mifflin Company, 1960), 640–42. See Clinton P. Anderson's review of *Atomic Diplomacy*, in *New York Times Book Review* (July 18, 1965), 24. See also, Henry L. Stimson, "The Bomb and the Opportunity," *Harper's Magazine* 192 (March 1946): 204.

The best study of Szilard is Carol S. Gruber, "Manhattan Project Maverick: The Case of Leo Szilard," *Prologue* 15 (Summer 1983): 73–87.

8. Harry S. Truman, *Memoirs: Years of Decision* (Garden City, N.Y.: Doubleday and Company, 1945), 416.

9. George Zhukov, *The Memoirs of Marshal Zhukov* (New York, 1971), 374–75, as cited in Martin J. Sherwin, "The Atomic Bomb and the Origins of the Cold War: U.S. Atomic-Energy Policy and Diplomacy, 1941–1945," *American Historical Review* 78 (October 1973), 956n.

10. Truman to James L. Cate, January 12, 1953, copy in Wesley Frank Craven and J. L. Cate, eds., *The Army Air Forces in World War II*. Volume 5: *The Pacific: Matterhorn to Nagasaki, June 1944 to August 1945* (Chicago: University of Chicago Press, 1953), opposite 712. Stalin's biographers also disagree on this issue. Isaac Deutscher felt it probable that Stalin knew more than he let on at Potsdam, while Adam Ulam has concluded that neither Stalin nor the Russian general staff

realized that Truman's reference was to an entirely new type of nuclear weapon. Isaac Deutscher, *Stalin: A Political Biography* (New York: Vintage, 1949, 1960), 547–48; Adam B. Ulam, *Stalin: The Man and His Era* (New York: Viking, 1973), 624–26.

11. *Nedelia*, weekly supplement of *Izvestiia* (February 6, 1983); translations provided by Advanced International Studies Institute and Richard G. Robbins, Jr.

12. Joint Committee on Atomic Energy, *Soviet Atomic Espionage* (Washington, D.C.: Government Printing Office, 1951), 22–25.

13. *Soviet Atomic Espionage*, 155–57. Alan Moorehead, *The Traitors* (New York: Harper and Row, 1952; 1963), 105.

14. Charles Maier, "Revisionism and the Interpretation of Cold War Origins," *Perspectives in American History* 4 (1970): 324.

15. J. Robert Oppenheimer, *The Open Mind* (New York: Simon and Schuster, 1955), 11.

16. William L. Laurence, *Men and Atoms: The Discovery, the Uses and the Future of Atomic Energy* (New York: Simon and Schuster, 1959), 125.

17. *New York Times* (September 26, 1945).

18. Kazuo Kawai, "*Mokusatsu*, Japan's Response to the Potsdam Declaration," *Pacific Historical Review* 19 (November 1950): 409–14; William J. Coughlin, "The Great *Mokusatsu* Mistake," *Harper's* 206 (March 1953): 31–40.

19. See Barton J. Bernstein, "The Perils and Politics of Surrender: Ending the War with Japan and Avoiding the Third Atomic Bomb," *Pacific Historical Review* 46 (1977): 1–28.

20. For example, Oppenheimer once said that he was familiar with all the arguments, and he was convinced by none.

21. Dudley, as cited in the *Los Alamos Chronicle*, 1943–1983, "Los Alamos: The Lab, the Town, the People" (Fortieth Anniversary special issue).

22. Henry Stimson, "The Decision To Use the Atomic Bomb," *Harper's Magazine* (February 1947): 99–100; Margaret Truman, *Harry S. Truman* (New York: William Morrow and Company, 1973); Herbert Feis, *The Atomic Bomb and the End of World War II* (Princeton: Princeton University Press, 1966); Karl T. Compton, "If the Atomic Bomb Had Not Been Used," *Atlantic Monthly* 178 (December 1946): 54–56. See also Truman's letter in *Atlantic Monthly* 179 (February 1947): 29; Harry S. Truman, *Year of Decisions*, 415–23. Michael J. Yavenditti, "The American People and the Use of Atomic Bombs on Japan: The 1940s," *The Historian* 36 (1974): 233–34.

23. Herbert Feis, *Japan Subdued: The Atomic Bomb and the End of the War in the Pacific* (Princeton: Princeton University Press, 1961);

Samuel Eliot Morison, "Why Japan Surrendered," *Atlantic Monthly* 206 (October 1960): 41–47. The quotation is on p. 47.

24. Taro Takemi, "Remembrances of the War and the Bomb," *Journal of the American Medical Association* 250 (August 5, 1983): 619.

25. Cited in Juan A. Del Regato, "Arthur Holly Compton," *International Journal of Radiation Oncology and Biological Physics* 7 (1981): 1586.

26. P. M. S. Blackett, *Fear, War, and the Bomb: Military and Political Consequences of Atomic Energy* (New York 1949), as quoted in Bert Cochran, *Harry Truman and the Crisis Presidency* (New York: Funk and Wagnalls, 1973), 173.

27. Norman Cousins and Thomas Finletter, "A Beginning for Sanity," *Saturday Review of Literature* 29 (June 15, 1946): 7; Justus D. Doenecke, *Not to the Swift: The Old Isolationists in the Cold War Era* (Lewisburg, Pa.: Bucknell University Press, 1979), 66–67. Certainly the new secretary of state, James Byrnes, was suspicious of the Soviets from the beginning. "His mind is full of his problems with the coming meeting of the foreign ministers and he looks to having the presence of the bomb in his pocket, so to speak, as a great weapon to get through the thing he has," Stimson noted in his diary. After failing to convince James Byrnes of the danger of an arms race, Leo Szilard wrote: "I thought to myself how much better off the world might be had I been born in America and become influential in American politics, and had Byrnes been born in Hungary and studied physics." Spencer R. Weart and Gertrude Weiss Szilard, eds., *Leo Szilard: His Version of the Facts* (Cambridge, Mass.: MIT Press, 1978), 184–85. Byrnes's reply was to leave Szilard completely out of his memoirs. James F. Byrnes, *Speaking Frankly* (New York: Harper and Brothers, 1947).

28. Lewis L. Strauss, *Men and Decisions* (New York: Doubleday and Company, 1962), 189–95. Fletcher Knebel and Charles W. Bailey, *No High Ground* (New York: Bantam Books, 1960), 181.

29. William D. Leahy, *I Was There* (London: Victor Gollancz, 1950), 502, 514.

30. United States Strategic Bombing Survey, *Japan's Struggle to End the War* (Washington, D.C.: Government Printing Office, 1946), 10–13; Charles L. Mee, Jr., *Meeting at Potsdam* (New York: Dell Publishing Company, 1975), 203, 204.

31. Las Vegas *Daily Optic* (September 20, 1945), Scrapbook 71, New Mexico State Records Center and Archives, Santa Fe. Vannevar Bush claimed that the use of the atomic weapons clearly ended the war, and that it was useless to argue over how much they advanced the end. Bush, *Modern Arms and Free Men: A Discussion of the Role of Science*

in Preserving Democracy (New York: Simon and Schuster, 1949), 91–92.

32. Gabriel Kolko, *The Politics of War: The World and United States Foreign Policy, 1943–1945* (New York: Random House, 1968); William Appleman Williams, *The Tragedy of American Diplomacy* (Cleveland: World Publishing Company, 1959); Gar Alperovitz, *Atomic Diplomacy: Hiroshima and Potsdam* (New York: Vintage, 1965). See also Barton J. Bernstein, "Roosevelt, Truman, and the Atomic Bomb, 1941–1945: A Reinterpretation," *Popular Science Quarterly* (Spring 1975): 23–68; and Barton J. Bernstein and Allen J. Matusou, eds., *The Truman Administration: A Documentary History* (New York: Harper and Row, 1966), 1–45. For a severe critique of their position, see Robert James Maddox, *The New Left and the Origins of the Cold War* (Princeton: Princeton University Press, 1973).

33. Martin J. Sherwin, *A World Destroyed*, 198; Barton J. Bernstein, "The Quest for Security: American Foreign Policy and International Control of Atomic Energy, 1942–1946," *Journal of American History* 60 (March 1974): 1003–44. See also Barton J. Bernstein, "The Atomic Bomb and American Foreign Policy, 1941–1945: An Historiographical Controversy," *Peace and Change* 11 (Spring 1974): 1–16; and Kenneth Glazier, Jr., "The Decision to Use Atomic Weapons Against Hiroshima and Nagasaki," *Public Policy* 18 (Winter 1969): 463–516; Louis Morton, "The Decision To Use the Atomic Bomb," *Foreign Affairs* 35 (January 1957): 334–53; Alice Smith, "Behind the Decision to Use the Atomic Bomb," *Bulletin of the Atomic Scientists* 14 (October 1958): 288–312; Martin J. Sherwin, "The Atomic Bomb as History: An Essay Review," *Wisconsin Magazine of History* 53 (1969–70): 128–34; Martin J. Sherwin, "The Atomic Bomb and the Origins of the Cold War: U.S. Atomic-Energy Policy and Diplomacy, 1941–45," *American Historical Review* 78 (October 1973): 945–68.

34. Kai Erikson, review of *Hiroshima and Nagasaki, New York Times Book Review* (August 9, 1981): 24.

35. Martin J. Sherwin, "The Atomic Bomb and the Origins of the Cold War: U.S. Atomic-Energy Policy and Diplomacy, 1941–1945," 946.

36. Quoted in Glazier, "The Decision to Use Atomic Weapons Against Hiroshima and Nagasaki," 515.

37. Nat S. Finney, "How FDR Planned To Use the A-Bomb," *Look* 14 (March 14, 1950): 25.

38. Statement of General Groves, U.S. Congress, Senate, Committee Hearings, 79th Congress, Senate Library, Vol. 809, 1946, Special, 39–40.

39. Statement of John Simpson, U.S. Congress, Senate, Committee

Hearings, 79th Congress, Senate Library, Vol. 809, 1946, Special, 303–4; Finney, "How FDR Planned To Use the A-Bomb," 24–26.

40. Robert J. Donovan, *Conflict and Crisis: The Presidency of Harry S Truman, 1945–1948* (New York: W. W. Norton, 1977), 46.

41. Winston S. Churchill, *Memoirs of the Second World War*, abridgement by Denis Kelly (Boston: Houghton Mifflin Company, 1959), 981.

42. Groves, *Now It Can Be Told*, 81, 149.

43. Lee Bowen, *A History of the Air Force Atomic Energy Program, 1943–1953. Introduction and Chapter I (1959): Project Silverplate, 1943–1946.* Typescript, NAM.   See also the account by James Les Rowe, *Project W-47* (Livermore, Cal.: Jā A. Rō, 1978).

44. Richard F. Newcomb, *Abandon Ship: The Death of the U.S.S. Indianapolis* (New York: Henry Holt and Company, 1958), 29–45.

45. The Interim Committee consisted of Henry L. Stimson, George L. Harrison, James L. Byrnes, Ralph A. Bard, William L. Clayton, Vannevar Bush, Karl T. Compton, and James B. Conant. They also called in four scientists: Arthur H. Compton, Enrico Fermi, Ernest O. Lawrence, and J. R. Oppenheimer.

46. Fletcher Knebel and Charles W. Bailey, "The Fight Over the A-Bomb," *Look* (August 13, 1963): 19–23; Alice Kimball Smith, *A Peril and a Hope: The Scientists' Movement in America, 1945–1947* (Cambridge, Mass.: MIT Press, 1970) treats this story.

47. Interview with Mick Daly, January 23, 1982, Albuquerque. After the bomb was used, the church was horrified. On August 7, the Vatican City paper *L'Observatore Romano* attacked America for using the weapons and compared this decision to Leonardo da Vinci's withdrawing of his invention of the submarine because of his fear it would be misused. In the September 1945 *Catholic Worker*, Dorothy Day excoriated Truman for the dropping of the bombs.

48. Jonathan Daniels, *The Man of Independence* (Philadelphia: J. B. Lippincott Company, 1950), 280.

49. Cochran, *Harry Truman and the Crisis Presidency*, quoted p. 174.

50. Stimson Diaries, 52: 34–38 (Reel 9).

51. Wesley F. Craven and James L. Cate, *The Army Air Forces in World War II*, Vol. 5, 707.

52. Kenneth M. Glazier, Jr., "The Decision to Use Atomic Weapons Against Hiroshima and Nagasaki," 29; Harry S Truman, *Year of Decisions*, 420.

53. Norman F. Ramsey, "History of Project A," LANL document, copy at National Atomic Bomb Museum, Albuquerque.

54. Gordon Thomas and Max Morgan Witts, *Enola Gay* (New York: Pocket Books, 1978) tells this story.

55. "Bethe Recalls 'Old Days,'" Santa Fe *New Mexican* (June 20, 1954).

56. Cochran, *Harry Truman and the Crisis Presidency*, 175. Some of the these statements may be found in Lisle A. Rose, *Dubious Victory: The United States and the End of World War II* (Kent, Ohio: Kent State University Press, 1973), 363–66; Merle Miller, *Plain Speaking: An Oral Biography of Harry S. Truman* (New York: G. P. Putnam's Sons, 1973), 228.

57. Robert H. Ferrell, ed., *Off the Record: The Private Papers of Harry S. Truman* (New York: Harper and Row, 1980), 52–53.

58. Lincoln to A. G. Hodges, April 4, 1864, in John G. Nicolay and John Hay, eds., *Abraham Lincoln: Complete Works*, vol. 2 (New York: Century Company, 1894), 508–9.

Chapter Nine

1. Bob Campbell et al., "Field Testing," *Los Alamos Science* 4 (Winter/Spring 1983): 167; and "Comment" by John Manley, "Nuclear Data," ibid., 123.

2. The Reminiscences of Kenneth T. Bainbridge, 58–59, COHP.

3. John F. Moynihan, *Atomic Diary* (Newark, N.J.: Baton Publishing Company, 1946), 62–66; John Hersey, *Hiroshima* (New York: Alfred A. Knopf, 1946); Santa Fe *New Mexican* (August 5, 1945), Scrapbook 71, NMSRA; Albuquerque *Journal* (August 9, 1945); Daniel Lang, *From Hiroshima to the Moon: Chronicles of Life in the Atomic Age* (New York: Simon and Schuster, 1959), 39–50.

4. A gripping account of the first visitors to the two Japanese cities may be found in the letter from M. D. Willcutts to Vice Admiral Ross T. McIntyre, September 23,1945, in the Atomic Bomb folder, FDR; Santa Fe *New Mexican* (November 28, 1945).

5. Donald L. Collins, "Pictures from the Past: Journeys into Health Physics in the Manhattan District and Other Diverse Places," in Ronald L. Kathren and Paul L. Ziemer, eds., *Health Physics: A Backward Glance* (New York: Pergamon Press, 1980), 41. See also Barton C. Hacker, "Nuclear Warfare and Radiation Safety: The Opening Scenes," Chapter 5 (draft) of "Elements of Controversy: A History of Radiation Safety in the Nuclear Weapons Testing Program." Copy lent by author.

6. "New Mexico's Atomic Bomb Crater," *Life* (September 24, 1945): 27–30.

7. G. Millard Hunsley, "Big Saucer-like Crater Marks Site of Bomb

Test," Albuquerque *Journal* (September 11, 1945). Paul Aebersold, *July 16th Nuclear Explosion—Safety and Monitoring of Personnel,* LA-616.

8. Untitled typescript of 1945 Park Service report on Trinity, 6, Trinity Site Raw Data file, NPS, Santa Fe.

9. Statement by Bob Krohn in typescript of interview conducted by Edwin C. Bearss, June 1968, copy in NPS, Santa Fe.

10. Howard M. Blakeslee, "Party of Newsmen Inspects Scene Near Alamogordo" (AP story), Albuquerque *Journal* (September 12, 1945).

11. Interview with Louis Hempelmann, December 22, 1981.

12. Will Harrison, "Even Ant Life Vanished from Vast Range Where Atomic Bomb Tested," Santa Fe *New Mexican* (September 12, 1945).

13. Alamogordo Chamber of Commerce (A. P. Grider, president, and Fritz Heilbronn, secretary) to director, August 11, 1945, Planning File, NPS, Santa Fe.

14. Las Cruces *Sun-News,* August 19, 1945, Scrapbook 71, NMSRA.

15. Report on Proposed Atomic Bomb National Monument, December 20, 1945, Trinity Site, Raw Data file, NPS, Santa Fe.

16. Two maps dated November 29, 1945 in NPS file.

17. Albuquerque *Journal* (September 9, 1945).

18. Ibid. (September 11, 1945), Scrapbook 71, NMSRA.

19. Albuquerque *Journal* (September 18, 1945), ibid.

20. Albuquerque *Journal* (September 22, 1945).

21. A Master Plan for Trinity National Historic Site, New Mexico, September 15, 1970, Planning File, NPS, Santa Fe.

22. Santa Fe *New Mexican* (July 10, 1947); Santa Fe *New Mexican* (February 21, 1946), Scrapbook 71, NMSRA; Robert Patterson to Secretary of Interior, March 4, 1947, Planning File, NPS, Santa Fe.

23. Santa Fe *New Mexican* (April 3, 1952; April 4, 1952).

24. "Report of Field Conference at Trinity Site," July 27, 1952, Department of Energy Coordination and Information Center, Las Vegas, Nevada, document 30087; "Trinity Site—Status of and Study on What to Do," August 26, 1952, Las Vegas, document 30090; Thomas L. Shipman to George P. Kraker, January 24, 1952, Las Vegas, document 30093.

25. Ken Tapman to Allan S. Kerr, July 18, 1973, in Trinity Site folder, NPS, Santa Fe; Joseph G. Rumburg to Superintendent, WHSA, April 8, 1974, Folder: Historic Sites and Structures Management and Preservation—Trinity Site, NPS, Santa Fe.

26. Joseph G. Rumburg to Superintendent, WHSA, April 8, 1974, Folder: Historic Sites and Structures Management and Preservation—Trinity Site; Carcie C. Clifford, Jr., to Commander, U.S. Army Materiel Command, September 11, 1973; Carcie C. Clifford, Jr., to Headquarters, Department of the Army, July 18, 1974; John F. Turnay to Regional Director, September 12, 1973, Folder Trinity Site, NPS, Santa Fe; Charles

D. Blackwell to Dean D. Meyer, "Preliminary Report of Surveys and Cleanup Program at Trinity Site from March 14, 1967 through April 5, 1967," April 17, 1967, Trinity Site Raw Data File, NPS, Santa Fe.

27. "Sleeping Beauty Awakens," *The Atom* 4 (May 1967): 9–11; interview with Norris Bradbury, July 30, 1982, Los Alamos, N.M.

28. Frederic L. Fey, Jr., *Health Physics Survey of Trinity Site* (LASL, 1967), LA-3719.

29. Thomas E. Hakonson and LaMar J. Johnson, "Distribution of Environmental Plutonium in the Trinity Site Ecosystem after 27 Years," Las Vegas, Nevada, document 14366; Richard L. Douglas, "Levels and Distribution of Environmental Plutonium Around the Trinity Site," October 1978, Las Vegas, document 13416.

30. Memorandum on National Survey of Historic Sites and Buildings: Atomic Energy Sites, April 19, 1965, Historic Landmark File, NPS, Santa Fe; Regional Director, Southwest Region to Chief, Office of Park Planning and Environmental Quality, WASO, April 11, 1980, contains a survey of these Park Service efforts, NPS, Santa Fe.

31. A Master Plan for Trinity National Historic Site, New Mexico, September 15, 1970, Planning File, NPS, Santa Fe.

32. *Empire Magazine,* Denver *Post* (July 6, 1967).

33. Interview with Norris Bradbury, July 30, 1982.

34. To: Chairman, Land and Water Conservation Fund Policy Group, From: Director, Heritage Conservation and Recreation Service, October 26, 1979, Planning File, Trinity Site, NPS, Santa Fe.

35. Interview with Robert W. Henderson, August 24, 1982, Albuquerque; interview with Roy W. Carlson (telephone), September 19, 1982.

36. Preston K. Pond to Norman R. Bandz, June 27, 1960, SHS.

37. Socorro *El Defensor-Chieftain* (February 12, 1965).

38. Major George C. Valentine to Socorro Chamber of Commerce, January 12, 1961; K. F. Hertford to Lt. Colonel Donal B. Jones, September 23, 1960, SHS.

39. *LASL News* (March 23, 1961): 6–7; San Diego *Union* (August 27, 1961). Copies at SHS.

40. Anderson read Simons's account into the *Congressional Record,* 1961, Appendix A 4089; copy at SHS.

41. Socorro *El Defensor-Chieftain* (November 19, 1964).

42. Interview with Edwin C. Bears (typescript), NPS, Santa Fe; "Jumbo Finally Getting Place in Atomic 'Sun,'" undated clipping, NAM.

43. "Trinity Site: *Open House Tour*" (White Sands Missile Range, 1982). See also the account by William L. Laurence, "Alamogordo, Mon Amour," *Esquire* (May 1965): 120–38.

44. Albuquerque *Journal* (August 6, 1947).

Epilogue

1. Robert Coughlan, "Dr. Edward Teller's Magnificent Obsession," *Life* (September 6, 1954): 65.

2. Alice Kimball Smith, *A Peril and a Hope: The Scientists' Movement in America, 1945–47* (Cambridge, Mass.: MIT Press, 1971).

3. Cited in Daniel J. Kevles, *The Physicists: The History of a Scientific Community in Modern America* (New York: Alfred A. Knopf, 1978), 375–76.

4. Mario Balibrera interview with Robert Bacher (1980), typescript, LANL.

5. Oppenheimer, cited in *Los Alamos Science* [Fortieth Anniversary issue] 4 (Winter/Spring, 1983): 25.

6. J. Robert Oppenheimer, "Physics in the Contemporary World," *Bulletin of the Atomic Scientists* 4 (1948): 66. See also Richard Rhodes, "I Am Become Death: The Agony of J. Robert Oppenheimer," *American Heritage* 28 (October 1972): quoted 78.

7. Freeman Dyson, *Disturbing the Universe* (New York: Harper and Row, 1939), 52–53.

8. Agnew in the New Mexico *Sun* (September 26, 1954).

9. Cited in John Shunny, "Not One to Freeze," *Century* 3 (July 6, 1983): 23.

10. Nigel Calder, *Nuclear Nightmares, an Investigation into Possible Wars* (London: British Broadcasting Corporation, 1979), 161.

11. Albuquerque *Journal* (April 15, 1983).

12. Alexander Babyonyshev, ed., *On Sakharov* (New York: Vintage, 1982).

13. I. I. Rabi, "How Well We Meant," April 14, 1983, tape, LAHS. See also, Gregg Herken, "Mad About the Bomb," *Harper's* 267 (December 1983): 49–55 for an interesting account of the fortieth reunion.

14. Interview with Frank DiLuzio, July 27, 1983, Santa Fe.

15. Thomas Powers, "Seeing the Light of Armageddon," *Rolling Stone* (April 29, 1982): 16.

16. Rabi, "How Well We Meant."

17. Santa Fe *New Mexican* (April 19, 1983).

Bibliography

Unpublished Materials

(The prefix LA indicates a Los Alamos report; DEL means that parts have been deleted. These are available at the J. Robert Oppenheimer Library, Los Alamos.)

Aebersold, Paul. *July 16th Nuclear Explosion—Safety and Monitoring of Personnel.* LA-616.

Anderson, Herbert L. *Radioactive Measurements at the 100 Ton Trial.* LA-282 (May 25, 1945).

Bainbridge, Kenneth T. *The Memoirs of Kenneth T. Bainbridge.* Columbia Oral History Program, Columbia University, New York.

———, ed. *Weapon Data: Effects and Instruments.* LASL Microfiche 1024 (October 27, 1947).

Blair, J. M., D. H. Frisch, and S. Katcoff (and J. M. Hubbard). *Detection of Nuclear Explosion Dust in the Atmosphere.* LA-418.

Bowen, Lee, *A History of the Air Force Atomic Energy Program, Introduction and Chapter I (1959): Project Silverplate, 1943–1946.* Typescript, NAM.

Chambers, Marjorie Bell. "Technically Sweet Los Alamos: The Development of a Federally Sponsored Scientific Community." Ph.D. dissertation, University of New Mexico, 1974.

Hacker, Barton C. "Elements of Controversy: A History of Radiation Safety in the Nuclear Weapons Testing Program." Manuscript draft. Lent by author.

Hempelmann, Louis H. *History of the Health Group.* (A-6) (March 1943–November 1945), 314. 7-Los Alamos.

———. *Preparation and Operational Plan of Medical Group (TR-7) for Nuclear Explosion, 16 July 1945.* LA-631.

Hoffman, Joseph G. *Nuclear Explosion, 16 July 1945: Health Physics Report on Radioactive Contamination throughout New Mexico Following the Nuclear Explosion.* LA-626.

Hubbard, Jack M. *Journal of J. Hubbard, Meteorologist,* LANL.

Larson, K. H., et al. *Alpha Activity Due to the 1945 Atomic Bomb Detonation at Trinity, Alamogordo, New Mexico.* UCLA-108 (1951).

———. *The 1949 and 1950 Radiological Soil Survey of Fission Product Contamination and Some Soil-Plant Interrelationships of Areas in New Mexico Affected by the First Atomic Bomb Detonation.* UCLA-108 (1951).

Larson, Kermit H. *Continental Close-in Fallout: Its History, Measurement and Characteristics* (typescript report). *Radi-ecology of Plutonium and Other Radionuclides in Desert Ecosystems,* edited by Paul Dunaway, et al., Report No. NVO 224. Las Vegas, Nevada: U.S. Dept. of Energy, Forthcoming.

Mack, J. E. *Weapon Data Effects and Instruments.* LA-1024 (1947).

The 1948 Radiological and Biological Survey of Areas in New Mexico Affected by the First Atomic Bomb Detonation. UCLA-32 (1949).

Stimson, Henry Lewis. *Diaries.* Microfilm edition, reel 9. Manuscripts and Archives, Yale University Library, New Haven, Connecticut.

Trinity Survey Program. UCLA-22 (1947).

Warren, Stafford L. *The Reminiscences of Stafford Warren.* UCLA Oral History Program, University of California at Los Angeles.

Published Works

Allardice, Corbin, and Edward R. Trapnell. *The First Reactor* (pamphlet). Washington, D.C.: Atomic Emergy Commission, 1957.

Allison, S. K. "A Tribute to Enrico Fermi." *Physics Today* 8 (January 1955): 9–10.

Alperovitz, Gar. *Atomic Diplomacy: Hiroshima and Potsdam.* New York: Vintage, 1965.

Badash, Lawrence. "Enrico Fermi and the Development of Nuclear Energy." In *Technology in America* by Carroll W. Pursell, Jr. Cambridge, Mass.: MIT Press, 1981.

Badash, Lawrence, Joseph O. Hirschfelder, and Herbert P. Broida, eds. *Reminiscences of Los Alamos, 1943–1945.* Dordrecht, Holland: D. Reidel Publishing Company, 1980.

Bainbridge, Kenneth T. *Trinity.* Los Alamos: Los Alamos Scientific Laboratory, 1976.

————. "A Foul and Awesome Display." *Bulletin of Atomic Scientists* 31 (May 1975).

————. "Prelude to Trinity." *Bulletin of the Atomic Scientists* 31 (April 1975).

Bar-Zohar, Michel. *The Hunt for German Scientists.* New York: Hawthorne Books, 1967.

Baxter, James P. *Scientists Against Time.* Boston: Little, Brown and Company, 1946.

Bernstein, Barton J. "The Atomic Bomb and American Foreign Policy, 1941–1945: An Historiographic Controversy." *Peace and Change* 11 (Spring 1974): 1–16.

————. "The Perils and Politics of Surrender: Ending the War with Japan and Avoiding the Third Atomic Bomb," *Pacific Historical Review* 46 (1977): 1–28.

————. "The Quest for Security: American Foreign Policy and International Control of Atomic Energy, 1942–1946." *Journal of American History* 60 (March 1974).

————. "Roosevelt, Truman, and the Atomic Bomb, 1941–1945: A Reinterpretation." *Political Science Quarterly* (Spring 1975): 23–68.

Bernstein, Barton J., and Allen J. Matusou, eds. *The Truman Administration: A Documentary History.* New York: Harper and Row, 1966.

Bernstein, Jeremy. "Profile of I. I. Rabi." *New Yorker* (October 21, 1945).

Bethe, Hans. "Can Air or Water Be Exploded?" *Bulletin of the Atomic Scientists* 1 (March 15, 1946).

Beyerchen, Alan D. *Scientists under Hitler.* New Haven: Yale University Press, 1977.

Blair, Henry A., ed. *Biological Effects of External Radiation.* New York: McGraw-Hill, 1954.

Blumberg, Stanley A., and Gwinn Owens. *Energy and Conflict: The Life and Times of Edward Teller.* New York: G. P. Putnam's Sons, 1976.

Bohr, Aage. "The War Years and the Prospects Raised by the Atomic Weapons." In *Niels Bohr,* edited by S. Rozenthal. Amsterdam: North-Holland Publishing Company, 1967.

Bolton, Herbert E. *Coronado on the Turquoise Trail: Knight of Pueblos and Plains.* Albuquerque: University of New Mexico Press, 1949.

Brode, Bernice. "Life at Los Alamos, 1943–45." *Atomic Scientists' Journal* 3 (November 1953).

————. "Tales of Los Alamos." *LASL Community News* (June 2–September 22, 1960).

Buck, Pearl S. "The Bomb—The End of the World." *American Weekly* (March 8, 1959): 9–12.

Bundy, Harvey H. "Remembered Words." *Atlantic Monthly* 199 (March 1957).

Burns, James MacGregor. *Roosevelt: The Soldier of Freedom*. New York: Harcourt Brace Jovanovich, 1970.

Bush, Vannevar. *Modern Arms and Free Men: A Discussion of the Role of Science in Preserving Democracy*. New York: Simon and Schuster, 1949.

————. *Pieces of the Action*. New York: William Morrow and Company, 1970.

Byrnes, James F. *Speaking Frankly*. New York: Harper and Brothers, 1947.

Cahn, Robert. "Behind the First Atomic Bomb." *Saturday Evening Post* (July 16, 1960).

Calder, Nigel. *Nuclear Nightmares: An Investigation into Possible Wars*. London: British Broadcasting Corporation, 1979.

Cave Brown, Anthony, and Charles B. MacDonald, eds. *The Secret History of the Atomic Bomb*. New York: Dial Press, 1977.

Churchill, Winston S. *Memoirs of the Second World War*. Boston: Houghton Mifflin Company, 1959.

Clark, Ronald W. *Einstein: The Life and Times*. New York: Avon Books, 1971.

————. *The Scientific Breakthrough*. New York: G. P. Putnam's Sons, 1974.

Cochran, Bert. *Harry Truman and the Crisis Presidency*. New York: Funk and Wagnalls, 1973.

Compton, Karl T. "If the Atomic Bomb Had Not Been Used." *Atlantic Monthly* 178 (December 1946): 54–56.

Conant, James B. *My Several Lives: Memoirs of a Social Inventor*. New York: Harper and Row, 1970.

Cornell, James. *The Great International Disaster Book*. New York: Pocket Books, 1976.

Coughlin, William J. "The Great *Mokusatsu* Mistake." *Harper's* 206 (March 1953): 31–40.

Cousins, Norman, and Thomas Finletter. "A Beginning for Sanity." *Saturday Review of Literature* 29 (June 15, 1946).

Craven, Wesley Frank, and J. L. Cate, eds. *The Army Air Forces in World War II*. Volume 5: *The Pacific: Matterhorn to Nagasaki, June 1944 to August 1945*. Chicago: University of Chicago Press, 1953.

Dabney, Thomas Ewing. "Red Hair of Hereford Cattle in Region Surrounding Site of Atomic Bomb Test in New Mexico Turns White." *New Mexico Stockman* 10 (November 1945).

Daniels, Jonathan. *The Man of Independence*. Philadelphia: J. B. Lippincott Company, 1950.

Davis, Noel Pharr. *Lawrence and Oppenheimer.* New York: Simon and Schuster, 1968.

Del Regato, Juan A. "Arthur Holly Compton." *International Journal of Radiation Oncology and Biological Physics* 7 (1981).

Deutscher, Isaac. *Stalin: A Political Biography.* New York: Vintage, 1949; 1960.

Dils, Lenore. *Horny Toad Man.* El Paso: Boots and Saddle Press, 1966.

Doenecke, Justus D. *Not to the Swift: The Old Isolationists in the Cold War Era.* Lewisburg, Pa.: Bucknell University Press, 1979.

Donovan, Robert J. *Conflict and Crisis: The Presidency of Harry S. Truman, 1945–1948.* New York: W. W. Norton, 1977.

Dudley, H. D. "The Ultimate Catastrophe." *Bulletin of the Atomic Scientists* 31 (November 1975).

Dupree, A. Hunter. "The *Great Instauration* of 1940: The Organization of Scientific Research for War." In *The Twentieth-Century Sciences: Studies in the Biography of Ideas,* edited by Gerald Holton. New York: W. W. Norton, 1970.

Dyson, Freeman. *Disturbing the Universe.* New York: Harper and Row, 1979.

Espinosa, J. Manuel, trans. *First Expedition of Vargas into New Mexico, 1692.* Albuquerque: University of New Mexico Press, 1940.

Evans, Robley D. "Radium Poisoning: A Review of Present Knowledge." *American Journal of Public Health* 23 (October 1933): 1017–73.

Feis, Herbert. *The Atomic Bomb and the End of World War II.* Princeton: Princeton University Press, 1961; 1966.

———. "The Secret That Traveled to Potsdam." *Foreign Affairs* 38 (January 1960): 300–317.

Fermi, Enrico. "The Development of the First Chain Reaction Pile." *Proceedings of the American Philosophical Society* 90 (January 1946): 20–24.

———. "Elementary Theory of the Chain-reacting Pile." *Science* 105 (January 1947): 27–32.

Fermi, Laura. *Atoms in the Family: My Life with Enrico Fermi.* Chicago: University of Chicago Press, 1954.

———. "Bombs or Reactors." *Bulletin of the Atomic Scientists* 26 (June 1970): 28–29.

———. *Illustrious Immigrants: The Intellectual Migration from Europe, 1930–1941.* Chicago: University of Chicago Press, 1968.

Ferrell, Robert H., ed. *Off the Record: The Private Papers of Harry S. Truman.* New York: Harper and Row, 1980.

Fine, Lenore, and Jesse A. Remington. *United States Army in World War II: The Technical Services: The Corps of Engineers: Construc-*

tion in the United States. Washington, D.C.: Government Printing Office, 1972.

Finney, Nat S. "How F.D.R. Planned to Use the A-Bomb." *Look* 14 (March 14, 1950).

Fitzpatrick, George. "Atomic Bomb Sight-Seeing." *New Mexico* 24 (January 1946).

Fleming, Donald, and Bernard Bailyn, eds. *The Intellectual Migration: Europe and America, 1930–1960.* Cambridge, Mass.: Harvard University Press, 1969.

Frisch, Otto R. "How It All Began." *Physics Today* 20 (November 1967): 43–52.

———. *What Little I Remember.* Cambridge: Cambridge University Press, 1979.

Glasstone, Samuel, and Philip T. Colan. *The Effects of Nuclear Weapons.* 3d ed. Washington, D.C.: Government Printing Office, 1977.

Glazier, Kenneth, Jr. "The Decision to Use Atomic Weapons Against Hiroshima and Nagasaki." *Public Policy* 18 (Winter 1969): 463–516.

Goodchild, Peter. *J. Robert Oppenheimer: Shatterer of Worlds.* Boston: Houghton Mifflin Company, 1981.

Goudsmit, Samuel A. *Alsos.* New York: Henry Schuman, 1947.

Gowing, Margaret. *Independence and Deterrence: Britain and Atomic Energy, 1945–1952.* Vol. 1: *Policy Making.* London: Macmillan Press, 1974.

Gregg, Josiah. *Commerce of the Prairies.* Edited by Max L. Moorhead. Norman: University of Oklahoma Press, 1954.

Groueff, Stephane. *Manhattan Project: The Untold Story of the Making of the Atomic Bomb.* Boston: Little, Brown and Company, 1967.

Groves, Leslie R. "Development of the Atomic Bomb." *Military Engineer* 38 (June 1946): 233–43.

———. *Now It Can Be Told: The Story of the Manhattan Project.* London: André Deutsch, 1963.

———. "Some Recollections of July 16, 1945." *Bulletin of the Atomic Scientists* 26 (June 1970): 21–27.

———. "The Story of the Atomic Bomb." *Think* 11 (November 1945): 5–7.

Gruber, Carol S. "Manhattan Project Maverick: The Case of Leo Szilard." *Prologue* 15 (Summer 1983): 73–87.

Hahn, Otto. "The Discovery of Fission." *Scientific American* 198 (February 1958): 76–84.

Hammond, George P., and Agapito Rey. *Don Juan de Oñate: Colonizer of New Mexico, 1595–1628.* Albuquerque: University of New Mexico Press, 1953.

Hammond, John Fox. *A Surgeon's Report on Socorro, N.M.: 1852*. Santa Fe: Stage Coach Press, 1966.

Hawkins, David. *Manhattan District History: Project Y, The Los Alamos Project*. Los Alamos: Los Alamos Scientific Laboratory, 1961.

Heisenberg, W. "Research in Germany on the Technical Application of Atomic Energy." *Nature* 160 (August 16, 1947): 211–15.

Heitzler, Gretchen. *Meanwhile, Back at the Ranch*. Albuquerque: Hidden Valley Press, 1980.

Herken, Gregg. "Mad About the Bomb." *Harper's* 267 (December 1983).

Hersey, John. *Hiroshima*. New York: Alfred A. Knopf, 1946.

Hewlett, Richard G., and Oscar E. Anderson, Jr. *The New World, 1939–1947: A History of the United States Atomic Energy Commission*. University Park, Pa.: Penn State University Press, 1962.

Hoffman, Frederic de. "Pure Science in the Service of Wartime Technology." *Bulletin of the Atomic Scientists* 31 (January 1975): 41–44.

Hyde, H. Montgomery. *The Atom Bomb Spies*. London: Hamish Hamilton, 1980.

In the Matter of J. Robert Oppenheimer. Washington, D.C.: Government Printing Office, 1954.

Irving, David. *The German Atomic Bomb: The History of Nuclear Research in Nazi Germany*. New York: Simon and Schuster, 1967.

Jauncey, G. E. M. "The Early Years of Radioactivity." *American Journal of Physics* 14 (July–August 1946): 226–41.

Jette, Eleanor. *Inside Box 1633*. Los Alamos: Los Alamos Historical Society, 1977.

Johnson, Charles W., and Charles O. Jackson. *City Behind a Fence: Oak Ridge, Tennessee, 1942–1946*. Knoxville: University of Tennessee Press, 1981.

Joint Committee on Atomic Energy. *Soviet Atomic Espionage*. Washington, D.C.: Government Printing Office, 1951.

Jungk, Robert. *Brighter Than a Thousand Suns: A Personal History of the Atomic Scientists*. New York: Harcourt Brace Jovanovich, 1958.

Kathren, Ronald L., and Paul L. Ziemer, eds. *Health Physics: A Backward Glance*. New York: Pergamon Press, 1980.

Kawai, Kazuo. "*Mokusatsu*, Japan's Response to the Potsdam Declaration." *Pacific Historical Review* 19 (November 1950): 409–14.

Kevles, Daniel J. *The Physicists*. New York: Alfred A. Knopf, 1978.

Kinzel, Marie. "The Town of Beginning Again." *Survey Graphic* 35 (October 1946): 354–57.

Kistiakowsky, George B. "Trinity—A Reminiscence." *Bulletin of the Atomic Scientists* (June 1980): 19–22.

Knebel, Fletcher, and Charles W. Bailey. "The Fight Over the A-Bomb." *Look* (August 13, 1963): 19–23.

————. *No High Ground.* New York: Bantam Books, 1960.

Kolko, Gabriel. *The Politics of War: The World and United States Foreign Policy, 1943–1945.* New York: Random House, 1968.

Koop, Theodore F. *Weapon of Silence.* Chicago: University of Chicago Press, 1946.

Kunetka, James W. *City of Fire: Los Alamos and the Atomic Age, 1943– 1945.* Rev. ed. Albuquerque: University of New Mexico Press, 1979.

————. *Oppenheimer: The Years of Risk.* Englewood Cliffs: Prentice-Hall, 1982.

Lamont, Lansing. *Day of Trinity.* New York: Atheneum, 1965.

Lang, Daniel. *From Hiroshima to the Moon: Chronicles of Life in the Atomic Age.* New York: Simon and Schuster, 1959.

Laurence, William L. "Address." In *The Beginnings of the Nuclear Age.* New York: Newcomen Society in North America, 1969.

————. "Alamogordo, Mon Amour." *Esquire* (May 1965): 120–38.

————. "The Atom Gives Up." *Saturday Evening Post* (September 7, 1940).

————. *Dawn Over Zero: The Story of the Atomic Bomb.* New York: Alfred A. Knopf, 1946.

————. *Men and Atoms: The Discovery, the Uses and the Future of Atomic Energy.* New York: Simon and Schuster, 1959.

Lawrence, Ernest O. "Thoughts by E. O. Lawrence." In *Foreign Relations of the United States, Diplomatic Papers: The Conference of Berlin (the Potsdam Conference),* 1945, vol. 2 of 2, 1369–70. Washington, D.C.: Government Printing Office, 1967.

Leahy, William D. *I Was There.* London: Victor Gollancz, 1950.

Leet, L. Don. "Earth Motion from the Atomic Bomb Test." *American Scientist* 34 (April 1946): 198–211.

Lewis, Richard S., and James Wilson (with Eugene Rabinowitch), eds. *Alamogordo Plus Twenty-five Years.* New York: The Viking Press, 1970.

Libby, Leona Marshall. *Uranium People.* New York: Charles Scribner's Sons, 1979.

Los Alamos: Beginning of an Era, 1943–1945. Los Alamos: Los Alamos Scientific Laboratory, c. 1960.

Maag, Carl, and Steve Rohrer. *Project Trinity.* Washington, D.C.: Defense Nuclear Agency, 1983.

Maddox, Robert James. *The New Left and the Origins of the Cold War.* Princeton: Princeton University Press, 1973.

Maier, Charles. "Revisionism and the Interpretation of Cold War Origins." *Perspectives in American History* 4 (1970).

McKee, Robert E. *The Zia Company of Los Alamos: A History*. El Paso: Carl Hertzog, 1950.

McPhee, John. *The Curve of Binding Energy*. New York: Farrar, Straus and Giroux, 1974.

Mee, Charles L., Jr. *Meeting at Potsdam*. New York: Dell Publishing Company, 1975.

Miller, Merle. *Plain Speaking: An Oral Biography of Harry S. Truman*. New York: G. P. Putnam's Sons, 1973.

Moorehead, Alan. *The Traitors*. New York: Harper and Row, 1952; 1963.

Morison, Elting E. *Turmoil and Tradition: A Study of the Life and Times of Henry L. Stimson*. Boston: Houghton Mifflin Company, 1960.

Morison, Samuel Eliot. "Why Japan Surrendered." *Atlantic Monthly* 206 (October 1960): 41–47.

Moynahan, John F. *Atomic Diary*. Newark, N.J.: Barton Publishing Company, 1946.

Newcomb, Richard T. *Abandon Ship: The Death of the U.S.S. Indianapolis*. New York: Henry Holt and Company, 1958.

Nicolay, John G., and John Hay, eds. *Abraham Lincoln: Complete Works*. Vol. 2. New York: Century Company, 1894.

"New Mexico's Atomic Bomb Crater." *Life* (September 24, 1945): 27–30.

Oppenheimer, J. Robert. "Niels Bohr and Atomic Weapons." *New York Review of Books* (December 17, 1966).

——. *The Open Mind*. New York: Simon and Schuster, 1955.

——. "Physics in the Contemporary World." *Bulletin of the Atomic Scientists* 4 (1948): 65–67, 85.

Pash, Boris T. *The Alsos Mission*. New York: Award House, 1965.

Rabi, I. I. *Science: The Center of Culture*. Cleveland: New American Library, 1970.

Rabi, I. I., et al., eds. *Oppenheimer*. New York: Charles Scribner's Sons, 1969.

Rhodes, Richard. "I Am Become Death: The Agony of J. Robert Oppenheimer." *American Heritage* 28 (October 1972): 72–82.

Rose, Lisle A. *Dubious Victory: The United States and the End of World War II*. Kent, Ohio: Kent State University Press, 1973.

Rowe, James Les. *Project W-47*. Livermore, Cal.: Jā A. Rō, 1978.

Schell, Jonathan. *The Fate of the Earth*. New York: Alfred A. Knopf, 1982.

Segrè, Emilio. *Enrico Fermi: Physicist*. Chicago: University of Chicago Press, 1970.

Sharp, William D. "The New Jersey Radium Dial Painters: A Classic in Occupational Carcinogenesis." *Bulletin of the History of Medicine* 52 (Winter 1978): 560–70.

Sherwin, Martin J. "The Atomic Bomb as History: An Essay Review." *Wisconsin Magazine of History* 53 (1969–70): 128–34.

———. *A World Destroyed: The Atomic Bomb and the Grand Alliance.* New York: Alfred A. Knopf, 1975.

Simmons, Marc. "Full Story of Bloody Jornada del Muerto Would Take a Book." *New Mexico Independent* 86 (April 2, 1982).

"Sleeping Beauty Awakens." *The Atom* 4 (May 1967): 9–11.

Smith, Alice Kimball. "Behind the Decision to Use the Atomic Bomb." *Bulletin of the Atomic Scientists* 14 (October 1958): 288–312.

———. *A Peril and a Hope: The Scientists' Movement in America: 1945–1947.* Chicago: University of Chicago Press, 1965.

———. "The Elusive Dr. Szilard." *Harper's Magazine* 221 (July 1960): 77–86.

Smith, Alice Kimball, and Charles Weiner, eds. *Robert Oppenheimer Letters and Recollections.* Cambridge, Mass.: Harvard University Press, 1980.

Smith, Cyril S. "Metallurgy at Los Alamos, 1943–1945." *Metal Progress* 65 (May 1954).

———. "Plutonium Metallurgy at Los Alamos During 1943–45." In *The Metal Plutonium,* edited by A. S. Coffinberry and W. N. Miner. Chicago: University of Chicago Press, 1961.

Smyth, Henry De Wolf. *Atomic Energy for Military Purposes: The Official Report on the Development of the Atomic Bomb under the Auspices of the United States Government, 1940–1945.* Princeton: Princeton University Press, 1947.

Spence, Clark C. "The Cloud Crackers: Moments in the History of Rainmaking." *Journal of the West* 18 (October 1979).

Sternglass, Ernest J. "The Death of All Children." *Esquire* (September 1969): 1a–1d.

———. *Low-level Radiation.* London: Earth Island, 1973.

Stewart, Irvin. *Organizing Scientific Research for War: The Administrative History of the Office of Scientific Research and Development.* Boston: Little, Brown and Company, 1948.

Stimson, Henry L. "The Bomb and the Opportunity." *Harper's Magazine* 192 (March 1946): 204.

———. "The Decision to Use the Atomic Bomb." *Harper's Magazine* 194 (February 1947): 99–100.

Stone, Robert S. "Health Protection Activities of the Plutonium Project." *Proceedings of the American Philosophical Society* 90 (January 1946): 11–19.

————, ed. *Industrial Medicine on the Plutonium Project: Survey and Collected Papers.* New York: McGraw-Hill, 1951.

Strauss, Lewis L. *Men and Decisions.* Garden City, N.Y.: Doubleday and Company, 1962.

Sundt, M. Eugene, and W. E. Naumann. *M. M. Sundt Construction Co.: "From Small Beginnings . . ."* New York: Newcomen Society in North America, 1975.

Szasz, Ferenc M. *The Divided Mind of Protestant America, 1880–1930.* Tuscaloosa: University of Alabama Press, 1982.

Szilard, Leo. "A Personal History of the Atomic Bomb." *University of Chicago Roundtable* (September 25, 1945): 14–16.

Teller, Edward. *Energy from Heaven and Earth.* San Francisco: W. H. Freeman, 1979.

————. "The Work of Many People." *Science* 121 (February 1955): 267–72.

Teller, Edward, with Allen Brown. *The Legacy of Hiroshima.* Garden City, N.Y.: Doubleday and Company, 1962.

Thomas, Charles A. "Epoch in the Desert." *Montsanto Magazine* 24 (1945): 28–31.

Thomas, Gordon, and Max Morgan Witts. *Enola Gay.* New York: Pocket Books, 1978.

Thompson, Fritz. "Behind the Barbed Wire Barricade." *Impact,* Albuquerque *Journal* Magazine (October 16, 1982): 4–9.

Truman, Harry S. *Memoirs: Years of Decision.* Garden City, N.Y.: Doubleday and Company, 1955.

Truman, Margaret. *Harry S. Truman.* New York: William Morrow and Company, 1973.

Truslow, Edith C. *Manhattan District History: Non-scientific Aspects of Los Alamos Project Y, 1942 through 1946.* Los Alamos: Los Alamos Scientific Laboratory, 1973.

Ulam, Adam B. *Stalin: The Man and His Era.* New York: Viking, 1973.

Ulam, S. M. *Adventures of a Mathematician.* New York: Charles Scribner's Sons, 1976.

Ulam, S. M., et al. "John von Neumann, 1903–1957." In *The Intellectural Migration,* by Fleming and Bailyn.

United States Stragetic Bombing Survey. *Japan's Struggle to End the War.* Washington, D.C.: Government Printing Office, 1946.

Voegtlin, Carl, and Harold Hodge, eds. *Pharmacology and Toxicology of Uranium Compounds.* New York: McGraw-Hill, 1949.

Warren, Stafford L. *The Role of Radiology in the Development of the Atomic Bomb.* Washington, D.C.: Surgeon General's Office, 1966.

Weart, Spencer R., and Gertrude Weiss Szilard, eds. *Leo Szilard: His Version of the Facts.* Cambridge, Mass.: MIT Press, 1978.

Weart, Spencer R. "Scientists with a Secret." *Physics Today* 29 (February 1973): 23–30.

Webb, J. H. "The Fogging of Photographic Film by Radioactive Contaminants in Cardboard Packaging Materials." *Physical Review* 76 (August 1, 1949): 375–80.

Weber, Robert L. *Pioneers of Science: Nobel Prize Winners in Physics.* Bristol and London: Institute of Physics, 1980.

Weisskopf, Victor F. "On Avoiding Nuclear Holocaust." *Technology Review* 83 (October 1980): 28–35.

———. *Physics in the Twentieth Century: Selected Essays.* Cambridge, Mass.: MIT Press, 1972.

Wigner, Eugene P. "An Appreciation on the 60th Birthday of Edward Teller." In *Properties of Matter under Unusual Conditions*, edited by Hans Mark and Sidney Fernbach. New York: John Wiley and Sons, 1969.

———. *Symmetries and Reflections: Scientific Essays of Eugene P. Wigner.* Bloomington: Indiana University Press, 1967.

Wilkes, Dan. "The Story of the Los Alamos 'Campus.'" *California Monthly* (June 1950).

Williams, William Appleman. *The Tragedy of American Diplomacy.* Cleveland: World Publishing Company, 1959.

Wilson, Jane, ed. *All in Our Time: The Reminiscences of Twelve Nuclear Pioneers.* Chicago: Bulletin of the Atomic Scientists, 1975.

Yavenditti, Michael J. "The American People and the Use of Atomic Bombs on Japan: The 1940s." *Historian* 36 (1974): 233–34.

Index